World Energy Crisis

Recent Titles in the

CONTEMPORARY WORLD ISSUES

Series

CRIMINAL JUSTICE

Modern Piracy: A Reference Handbook David F. Marley

ENVIRONMENT

Climate Change: A Reference Handbook David I. Downie, Kate Brash, and Catherine Vaughan

GENDER AND ETHNICITY

Women and Crime: A Reference Handbook Judith A. Warner
Women in Developing Countries: A Reference Handbook Karen L. Kinnear

POLITICS, LAW, AND GOVERNMENT

Tax Reform: A Reference Handbook, Second Edition James John Jurinski

SCIENCE, TECHNOLOGY, AND MEDICINE

Medical Tourism: A Reference Handbook Kathy Stolley and Stephanie Watson
Virtual Lives: A Reference Handbook James D. Ivory

SOCIETY

Child Soldiers: A Reference Handbook David M. Rosen
World Sports: A Reference Handbook Maylon Hanold

Books in the **Contemporary World Issues** series address vital issues in today's society such as genetic engineering, pollution, and biodiversity. Written by professional writers, scholars, and nonacademic experts, these books are authoritative, clearly written, up-to-date, and objective. They provide a good starting point for research by high school and college students, scholars, and general readers as well as by legislators, businesspeople, activists, and others.

Each book, carefully organized and easy to use, contains an overview of the subject, a detailed chronology, biographical sketches, facts and data and/or documents and other primary source material, a directory of organizations and agencies, annotated lists of print and nonprint resources, and an index.

Readers of books in the **Contemporary World Issues** series will find the information they need in order to have a better understanding of the social, political, environmental, and economic issues facing the world today.

CONTEMPORARY WORLD ISSUES

Science, Technology, and Medicine

World Energy Crisis

A REFERENCE HANDBOOK

David E. Newton

Santa Barbara, California • Denver, Colorado • Oxford, England

Library of Congress Cataloging-in-Publication Data

Newton, David E.
World energy crisis : a reference handbook / David E. Newton.
p. cm. — (Contemporary world issues)
Includes bibliographical references and index.
ISBN 978-1-61069-147-5 (hbk. : alk. paper) —
ISBN 978-1-61069-148-2 (e-book) 1. Energy security. 2. Energy consumption. 3. Energy conservation. 4. Energy development. 5. Energy industries. 6. Energy policy. I. Title.
HD9502.A2N498 2012
333.79—dc23 2012016975

ISBN: 978-1-61069-147-5
EISBN: 978-1-61069-148-2

16 15 14 13 12 1 2 3 4 5

This book is also available on the World Wide Web as an eBook. Visit www.abc-clio.com for details.

ABC-CLIO, LLC
130 Cremona Drive, P.O. Box 1911
Santa Barbara, California 93116-1911

This book is printed on acid-free paper ∞

Manufactured in the United States of America

Contents

List of Tables and Figures

Tables

Figures

Preface

What does it mean to say that there is a world energy crisis today? On the one hand, it would appear that the world has a huge supply of the three fossil fuels, oil, coal, and natural gas. In its 2010 *Statistical Review of World Energy,* the BP oil company reported that proven global oil reserves at the end of 2010 amounted to 1,383 billion barrels, an increase of 25 percent over the 2000 reserves. Experts predict that those reserves are sufficient to meet the world's need for oil for as much as 80 years. The situation looks even more promising for natural gas, where proven reserves are expected to last at least 60 years, and coal reserves, up to about 160 years. So how can there be an energy crisis?

One reason for energy concerns is that consumption is increasing much more rapidly than is the discovery and the production of fossil fuels. This increase is a result of two factors, the first of which is population growth. It took 1,800 years for the world to reach a population of one billion, 123 more years for it to reach a population of two billion, and 33 years for it to reach three billion in 1960. The world has added an additional one billion people about every decade since then, passing the seven billion mark in late 2011. The more people there are on the Earth, the greater the demand for energy resources, such as coal, oil, and natural gas.

Energy demand is increasing also because of an improved standard of living for many people throughout the world. And an increased consumption of fossil fuels closely corresponds to a higher standard of living. That trend has not been so obvious in developed nations, where energy consumption per capita has remained relatively constant over the past two decades. But the use of energy has increased dramatically in many developing nations, where the use of fossil fuels has been a driving force in the improvement of people's lives. In Brazil, as an example, per capita energy consumption increased from 897 kgoe (kilograms of oil equivalent) in 1990 to 1,124 kgoe in 2005, an increase of 25 percent. Those statistics were even more dramatic for other larger developing nations, such as India (377 kgoe in 1990 to 491 kgoe in 2005; 30% increase), Indonesia (579 kgoe in 1990 to 814 kgoe in 2005; 43% increase), and China (760 kgoe in 1990 to 1,316 kgoe in 2005; 73% increase).

Concerns about a world energy crisis also reflect doubts about the amount of coal, oil, and natural gas that can eventually be recovered. While the quantity of proved reserves for the fossil fuels continues to inch up each year, those fuels are a nonrenewable resource. That is, coal, oil, and natural gas that were originally produced millions of years ago is all that the world has of those resources; new reserves are not being created. Ultimately, the world will run out of those supplies, although vigorous dispute exists as to how far into the future that event will occur. Many people believe that the era of peak oil, the maximum amount of oil produced in a single year, has already occurred, and the world is now facing a period when less and less oil (and coal and natural gas) will be available every year into the future.

Geopolitical factors also contribute to the world energy crisis. These factors result from the fact that the regions of the Earth where reserves of oil and natural gas (and, to a lesser extent, coal) are available are very different from the regions where they are most widely used. Five of the ten largest oil producing nations in the world are located in the Middle East, two are in Africa, and one is in South America; only two (Canada and Russia) are in the Northern Hemisphere. By contrast, most of the largest oil consuming nations in the world (such as China, Germany, India, Japan, and South Korea) have very limited natural sources of oil. Even developed nations with substantial oil reserves, such as the United States, may still be oil importers. In April 2011, for example, the United States imported 61 percent of its oil, although it had proved reserves of almost 31 billion barrels. With this dichotomy of reserves and consumption, oil (and to a lesser extent, coal and natural gas) have become political weapons in the interaction among countries around the world, an added element in the world's energy crisis.

Finally, the world faces an energy crisis because of a host of environmental issues created by the production and the consumption of coal, oil, and natural gas. The removal of these natural resources can create devastation of the landscape (as when mountain tops are removed to get at coal); additional problems may arise during transportation (as when oil tankers break apart and spill their contents in the ocean); and even more issues arise when the fossil fuels are burned (as occurs as a result of air pollution and possible global climate change). Appropriate ways in which these issues can be addressed have become crucial aspects of the world energy crisis.

One possible solution to this crisis involves the increased use of alternative forms of energy: solar, wind, hydroelectric, nuclear, geothermal, biomass, and the like. Humans have made use of these nonfossil-fuel energy sources for centuries, but, over most of that time, they have played a relatively minor role in the overall human energy equation. Many observers now believe that a dramatic revolution in the world's energy story is about to occur, with these renewable forms of energy assuming an even greater role in meeting the world's energy needs.

This book is intended as a research guide for further study of the world's energy crisis. The first two chapters provide background information about the world's energy crisis, whereas

Chapter 3 provides a sample of various expert views on that crisis. The remaining chapters offer resources to guide one's further study of the energy crisis: profiles of individuals and organizations related to energy issues; a chronology of events; a list of print and electronic resources; some related documents and data; and a glossary of essential terms.

World Energy Crisis

"WHAT WILL HE GROW TO?"

1 Background and History

Time: 80 BCE
Place: Northern Greece

Ctronius speaks to his daughter, Alverinda: Alverinda, my daughter. You do not have to go to the vineyard today. Your uncle Demosthenes has invented a new method for bringing water from the river to the grape vines. You no longer have to carry water two buckets at time from the river to the fields. Go instead to the village teacher, old Socatius, and ask if you can join his classes. Maybe his lessons will give you knowledge you can use to live a better life.

Energy in Antiquity

No one really knows for certain when humans first started using natural resources in the world around them to relieve them of the burdens of everyday life. At first, they had to rely on their own muscle power to carry water, to lift large rocks, and to move heavy loads for long distances. Construction of great monuments of early cultures, such as the pyramids in Egypt and the Great Wall of China, was accomplished only with human muscle power exerted by untold numbers of slaves, who often lost their lives in the process. But anthropologists know that early on in history, humans found ways to

This 19th-century illustration personifies the significance of the coal industry in the 1800s. (Library of Congress)

handle these chores more easily, using resources other than the power of human muscles. As early as 6000 BCE, for example, humans had domesticated a number of animals, including the camel, the ass, the horse, and the ox, as a means of transportation, providing a major improvement over walking or running. But it may have been another 4,000 years before humans began to using these animals for other purposes, such as pulling carts and wagons, providing expanded methods of transportation (Kubiszewski 2012).

By this time, humans had also begun to appreciate the value of certain natural resources in making their lives easier and more comfortable. As early as 3000 BCE, they had discovered that such natural products could be captured and burned to produce energy for a variety of purposes. Among the earliest of these discoveries occurred in China, where coal was first used as a fuel in about 3000 BCE. Much later, around 500 BCE, the Chinese also discovered a way to collect natural gas through bamboo pipes inserted into the ground, a fuel they used to heat their homes and used for other purposes. Petroleum was also in wide use as early as about 6000 BCE, although usually not as a fuel, but as a construction material or as a medicine ("History" 2012).

Resources that we now think of as alternative forms of energy—wind, solar, and tidal power, for example—were widely used by early humans. The first sailing ships, designed to make use of wind power, were probably built in Egypt in at least 3500 BCE, if not much earlier. Early humans also used solar power to heat their homes. By 400 BCE, for example, many Greek cities were designed and built in such a way that most homes faced south, providing a dependable source of heat during the cool winter months. Water power was developed as a source of energy in early cultures, where it was used to drive water wheels for the grinding of grain. In about 240 BCE, as an example, Greek inventor Ctesibius of Alexandria invented a double-action piston pump that was used to lift water from one level to another, relieving humans of the kind of onerous tasks described in the beginning of this chapter. Even tidal power, an energy resource only rarely used in the modern world, was already being employed as a reliable source of energy in some parts of northern Europe as early as the eighth century (Kubiszewski 2012).

The Great Transition: The Industrial Revolution

The way that most humans lived in the 17th century was probably not a great deal different from the way they had lived for the preceding thousand years or more. People still relied on animals, wind power (windmills), water power (water wheels), tidal power, solar power, and the power of their own muscles to carry out most of their daily tasks. About the only fuels available to them were wood or other forms of biomass. The term *biomass* refers to any form of organic matter. In energy discussions, it refers more specifically to organic matter that can be used as a fuel, in which case it is sometimes also known as *biofuel.* Historically, wood was used so widely as a fuel (as well as for other purposes, such as construction) that many regions had been denuded of forests early in history. Historians now believe that the loss of forests was a major factor in the fall of many early civilizations, including those of Sumeria, Assyria, and Babylonia, dating to the third millennium BCE. They also know that, as early as about 380 BCE, the natural forests of Greece had been so thoroughly destroyed that the government had to issue edicts to restrict any further lumbering. In Europe, the demand for wood was so great that by the early 16th century, the continent's forests had essentially been lost ("Sustainability in a Changing World" 2012; Oosthoek 2012).

Still, by 1700, wood and other forms of biomass were essentially the only fuel available to humans in Europe and other parts of the world. Only a few years into the 18th century, however, a dramatic change began to occur that revolutionized the way energy was produced and consumed by humans. That change is now known as the Industrial Revolution, a transformation that began in Great Britain in which many tasks previously performed by human or animal labor were now carried out by machines. One feature of the Industrial Revolution was the growing importance of coal as a source of energy. Although Great Britain, like the rest of Europe, was rapidly running out of the wood it needed to heat homes and to use in industrial operations, it had a critical backup fuel: coal. The nation had huge coal resources that it had never called on to any great extent because of the ready availability of wood. By the early 18th century, however, that situation had begun to change. In 1709, for example, ironmonger Abraham Darby invented a method for converting coal into coke for use as a fuel in the manufacture of iron pots. Traditionally, Darby would have used wood or charcoal for this purpose. (Charcoal is a fuel manufactured by heating wood in the absence of air. It is a more efficient fuel than wood itself.) However, wood was becoming much more expensive, and Darby had no access to the fuel for his iron works. Darby could not use coal in his forge either, since impurities in coal produced a type of iron that was too brittle to use. By trial-and-error experimentation, Darby found that he could convert coal to coke by heating the coal in the absence of air (as in the conversion of wood to charcoal). The coke produced by this method was a more efficient fuel than coal and also resulted in the production of a stronger type of iron pot (Derry and Williams 1993, 474–80).

Perhaps the defining moment of the early Industrial Revolution was the invention in 1712 by Thomas Newcomen of the steam engine. The steam engine, later improved by James Watt in 1769, used steam produced by boiling water to drive a piston that could perform mechanical work. The steam engine was of great significance to the coal industry for two major reasons. First, it soon revolutionized many operations previously carried out by hand, such as plowing, threshing, and weaving. It also completely changed traditional modes of transportation, with the first steamship invented in 1775 by Jacques Perrier and the first steam locomotive in 1804 by Richard Trevithick. Inventions like these created a huge demand for coal (Derry and Williams 1993, chapter 11).

The steam engine was also important because it provided a way to remove water from deep coal mines. As the demand for coal grew in Great Britain, deposits of the mineral at or close

The first steam engine designed and built in the United States. Steam engines provided a new source of energy in the 18th century that quickly became widespread in Europe and the United States. (Library of Congress)

to the surface were mined first. Those deposits were relatively sparse, however, and they were soon exhausted. Companies found that they had to dig deeper and deeper mines to gain access to the extensive underground coal resources in the country. However, a serious hazard involved in the construction of deep coal mines was the seepage of ground water, limiting mining operations or even making them impossible. Trevithick's invention, however, provided a relatively simple solution to the problem of mine seepage. Steam-powered pumps were used to remove water and allow mining at even deeper depths. This development allowed the construction of coal mines at depths of up to 2,500 feet by the end of the 19th century (Derry and Williams 1993, 324–26).

The importance of coal during the 18th century was enhanced by a series of discoveries that expanded the ways in which it could be used. The first of those discoveries was made in 1784 by French professor of philosophy Jean Pierre Minkelers. Minkelers, along with a number of other researchers, was looking for a gas that could be used in place of the relatively expensive hydrogen gas to lift balloons, the latest rage among inventors at the time. Minkelers found that he could produce a low-density gas by heating soft coal in the absence of air. When he also found a way to purify the gas, he began using it to illuminate his lecture room at the University of Louvain. The gas discovered by Minkelers was later to become known as coal gas, illumination gas, or town gas. It consists of a complex mixture of flammable elements and compounds, including methane, carbon monoxide, and hydrogen, as well as some nonflammable substances, such as carbon dioxide and nitrogen. The presence of the nonflammable substances reduces the efficiency with which coal gas burns and also accounts for its smoky flame ("Random Walks in the Low Countries" 2012).

No commercial application of Minkelers's discovery occurred until more than a decade later. Then, in 1792, Scottish mechanic William Murdock repeated Minkelers's experiments, producing enough coal gas to light his own home in Redruth, Cornwall. Six years later, while employed at the Soho Foundry in Birmingham, he had developed the process to a point where commercial lighting with the gas could be offered by the firm of Bolton & Watt. The first company to adopt the use of coal gas lighting for its buildings was a large cotton mill in Salford, which, in 1806, installed 900 gas lights to illuminate its factory. By the early 19th century, most commercial buildings and homes were lighted with some version of coal gas, as were many homes in Europe and the United States (Thomson 2003).

The demand for coal increased significantly at the end of the 19th century for a second reason: the growing requirement for coke in the young iron and steel business. Coal is used in the production of iron to extract the metal from its ores. But coal is not a suitable fuel for converting iron into steel, a far more desirable product commercially. The reason is that coal contains impurities such as sulfur and phosphorus that make steel weaker and more brittle. Substituting coke for coal in the steel-making process eliminates this problem, and by the end of World War I, most steel-making operations (88% in the United States) used coke rather than coal.

Because of its many and varied uses, coal radically changed the energy equation in Great Britain and the rest of Europe during the Industrial Revolution. (The term *energy equation* is often used in a somewhat ambiguous way to describe the mix of energy resources and economic and environmental factors at work in a nation, region, or other geopolitical unit.) By 1800, coal had taken over about a quarter of the energy market and more than three-quarters of the market a century later. Wood was displaced from the market by similar ratios, accounting for 80 percent of all the energy produced in Great Britain in 1800, but only 20 percent in 1900. The increasing importance of coal can also be seen in the absolute amounts of the fuel produced between 1700 and 1900. At the beginning of that period, in 1700, the nation produced only 3.0 million tons of coal, all of it from surface sources that required no underground mining. Fifty years later, as the Industrial Revolution was just getting under way, that total rose to 5.2 million tons. After another 50 years, in 1800, the nation produced twice as much coal (10 million tons), much of it now from underground mines. In the next two 50-year periods, the production of coal quintupled to 55 million tons in 1850 and then quintupled again to 275 million tons in 1900 ("Coal Mines in the Industrial Revolution" 2012).

A similar pattern was developing in the rest of Europe. By the end of the 19th century, the United States had surpassed Great Britain as the world's largest producer of coal, with an annual output of about 350 million tons. The world's third largest producer was Germany (121 million tons), followed by France (36 million tons), Belgium (22 million tons), Russia (19 million tons), and Austria (13 million tons). The leading producing states in Asia were Japan (10 million tons) and India

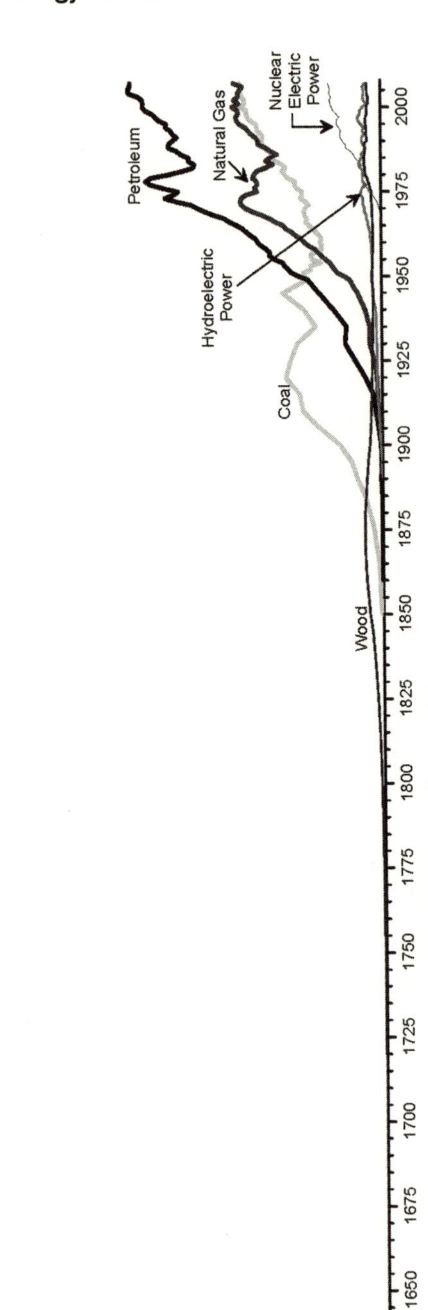

Figure 1.1 Primary Energy Consumption by Source, 1635–2008.

Source: U.S. Department of Energy, Energy Information Administration. Annual Energy Review, 2008, p. 7. Available online at http://www.scribd.com/doc/29251967/US-Annual-Energy-Review-2008.

(8 million tons); in Africa, Transvaal (2.5 million tons) and Natal (1 million tons), both parts of modern-day South Africa; and in the South Pacific, Australia (7 million tons) and New Zealand (1.6 million tons) ("Coal" 2012).

The changing balance between wood and coal was mirrored in many other parts of the world. As Figure 1.1 shows, in the United States the rise of coal and the fall of wood followed a pattern similar to that seen in Great Britain. When the nation was founded, the continent was essentially covered with forests, providing a seemingly unlimited supply of wood. And for the first 250 years of the nation's history, its citizens relied almost entirely on wood and charcoal for their heating and industrial needs. By 1875, however, the use of coal for heating and industrial purposes began to grow, and by the early 20th century it was the undisputed leading source of energy in the nation.

Birth of a Modern Giant: The Discovery of Oil

At the very time when coal was becoming king of energy sources in the late 19th century, a second form of energy was breaking upon the scene elsewhere in the world. Humans had been familiar with and used petroleum (crude oil) for many centuries. Since it can sometimes be found in above-ground pools, it is often readily accessible and had been used for oil lamps and other purposes as long as at least 5000 BCE. However, no serious efforts had been made to mine or otherwise collect the resource until the late 16th century (1594), when explorers dug what were probably the first real oil wells in the region of Baku in modern Azerbaijan. These wells were hand-dug to a depth of about 100 feet and produced a modest flow of crude oil (Aliyev 2012). (The term *crude oil* refers to a product removed from the earth, whereas the term *petroleum* refers both to crude oil and to any products made from crude oil.)

The first modern oil well was drilled on the Aspheron peninsula northeast of Baku in 1848 by Russian engineer F. N. Semyenov. By 1861, this well was producing 90 percent of all the crude oil being collected in the world. In 1854, Polish druggist Ignacy Lukasiewicz became interested in the possibility of using crude oil as a source of kerosene as a substitute for whale oil in lamps. He had heard of a method developed by Canadian geologist Abraham Gesner for distilling crude oil to obtain kerosene, and saw the potential for a huge profit in introducing the method to his native Poland. He and his colleagues began digging wells first 164 feet, and later 492 feet, in depth, successfully collecting ever and ever more pure forms of crude oil. By 1858, Lukasiewicz had constructed one of the world's first distillation plants to separate the desired kerosene from his crude oil products. Oil exploration continued apace in other parts of Eastern Europe at about the same time. Drilling efforts were especially successful in Romania, where the first written production records were collected in 1857, with a reported production of 275 tons of crude oil for that year. Cîmpina, Romania, was also the site of the world's first large oil refinery, which opened for operation in 1857 (Robinson 2012).

Explorers in North America were also becoming interested in the search for oil. The first discoveries of the fuel were made in 1858 in a region of southern Ontario around the towns of Oil Springs, Bothwell, and Petrolia (whose names all reflect their association with petroleum). The wells were extraordinarily productive, apparently having struck a vein of inexhaustible oil, as described by one observer of the time. Within the next decade, dozens of wells had been constructed, producing anywhere from 150 to 7,500 barrels of oil per day. Only a year after the Canadian strikes, the first successful oil well in the United States was drilled at Titusville, Pennsylvania, by Colonel Edwin Drake (who chose the sobriquet colonel to impress potential investors in his wells). Drake's discovery is sometimes claimed to be the first well drilled solely and specifically for the purpose of finding oil, although it is hardly the first oil well in the world, as is sometimes claimed.

Talking about Energy

Before proceeding with a review of the background and the history of worldwide energy issues, it will be worthwhile to take a moment to review the concept of energy as a scientific topic. Most people know and use the term *energy* as a way of describing one's physical, mental, and emotional states, as in: "I just didn't have the energy to take on that project." But the word *energy* has a very specific meaning in physics, and that meaning carries with it other related terms of some importance, such as work and power.

Energy is the capacity to do work. In science, the term *work* also has a very specific meaning. Work is the change that occurs when a force (a push or a pull) moves an object over some given distance. Algebraically, work is defined as:

$$W = f \times d$$

The standard unit for measuring work is the joule, which is the amount of work involved in moving a force of one newton across a distance of one meter. Other units in which work can be expressed include the erg, foot-pound, foot-poundal, liter-atmosphere, horsepower-hour, therm, BTU (British thermal unit), calorie, and Calorie (a thousand times the size of the small-c calorie).

So what does it mean to say that energy is the capacity to do work? Consider a rock resting at the top of a hill. If the rock begins to roll down the hill, it may come into contact with other objects: blades of grass, grasshoppers, or a bicycle. When the rock strikes the grass, the grasshopper, or the bicycle, it has the ability to move the object, to push down on the grass or the grasshopper, or to cause a dent (or worse) in the bicycle. In each case, the rock has moved the object (grass, grasshopper, or bicycle) over a distance. So the rock has energy, the ability to do work.

That energy may be of two different kinds, however. When sitting at the top of the hill, the rock can't do anything because it's not in motion. At this point, it has only *potential energy,* the capacity for moving objects once it is in motion. If the rock begins to actually move, it then has *kinetic energy,* or the energy of motion. At this point, blades of grass, grasshoppers, and bicycles are at risk of being moved because of the rock's movement.

Energy can be classified in another way also, according to the form that it takes. The rock sitting at the top of a hill has *gravitational energy* because it tends to move downward, toward the center of the Earth. A bolt of lightning has *electrical energy* because it consists of positively or negatively charged particles that have the capacity to do work. Chemical substances also have a form of energy known as *chemical energy.* Such substances are held together by chemical bonds, such as those that hold hydrogen atoms and oxygen atoms together in water. When those bonds break, they release some of that chemical energy.

One of the most important laws in physics is that one cannot create or destroy energy. Energy can change from one form into another form, but it never disappears, nor can it ever be created. For example, when you burn a gallon of gasoline, chemical bonds that hold carbon and hydrogen atoms together in gasoline molecules are broken. The chemical energy that is released changes into a different form, called heat (or *thermal energy*). And that simple statement summarizes almost all you really need to know about fuels. People burn wood, coal, oil, and gas in order to break the chemical bonds that hold carbon atoms and hydrogen atoms together in these substances, converting that energy to heat, which is used to boil water or perform other useful tasks.

One final term is worth knowing: power. Again, as with *energy* and *work,* the term *power* has a somewhat different meaning in science than in everyday life. In science, power is defined as the rate at which work is done. Think of two individuals picking up a box and placing it on the shelf. Both individuals do the same amount of work because they move the same object through the same distance. But one person moves the box more quickly than does the second. So the first person exerts more power than does the second. The formula for power use is:

$$P = \frac{W}{t}$$

The standard unit for measuring power is the watt (W), named after James Watt. One watt is equal to one joule per second (J/s). Other common units of power are the erg per second (erg/s), the horsepower (hp), the metric horsepower (Pferdestärke [PS] or the cheval vapeur [CV]), and the foot-pound per minute (ft-lb/min).

Energy Units

The most important part of this discussion has to do with the units in which energy is expressed. As you read about the production and consumption of energy, you will find references to some number of joules or kilowatt-hours or Btus or other units. It is helpful to know what those units mean. Begin by recalling that the basic unit for all forms of energy is the joule, represented by the letter J. A joule is a really small amount of energy. If a person weighing 60 kilograms (about 130 pounds) briskly climbs a single flight of stairs, he or she expends about 1800 J of energy. Or, as another way to think about the unit, burning a single liter of gasoline (a quarter of a gallon) releases about 40,000,000 J of thermal energy (heat). Because the joule is such a small unit of measure, a unit more commonly used is the kilojoule (kJ), or one thousand joules, or the megajoule (MJ), or one million joules.

For your future reference in this and later chapters, Table 1.1 shows conversions for various units of energy. These conversions are not exact, because various kinds of fuels differ in their energy contents. For example, the energy content of crude oil depends on the source from which it comes, and the energy content of coal depends on the type of coal (anthracite, bituminous, peat, etc.).

Table 1.1 Energy Unit Conversions

Fuel	Amount	Joules	Kilowatt-hours*	Btu
Coal	1 ton	28,000,000,000	7800	27,000,000
Oil	1 barrel†	6,100,000,000	1700	5,800,000
Natural gas	1 cubic foot	1,000,000	0.29	950
Wood	1 ton	10,000,000,000	2900	9,500,000
Uranium	1 gram	82,000,000,000	23,000	78,000,000

*Kilowatt-hours is most commonly used to measure electrical energy.
†1 barrel = 42 gallons.

Finally, a few concluding notes about units of measurement.

1. A watt is, like the joule, a relatively small quantity, and a much more common unit of power measurement is the kW, equal to a thousand watts, or a megawatt (MW), equal to a million watts. Power is also measured in the English system in horsepower (hp), a unit originally based on the amount of work a horse can do in a given period of time. One hp is equal to 745.7 W.
2. Discussions of energy issues often refer to very large quantities whose measurement may make use of metric prefixes indicating large numbers. In addition to the familiar kilo- (k), for one thousand, other indicators of very large numbers are mega- (M) for one million, giga- (G) for one billion, and tera- (T) for one trillion. An especially common unit of measurement in discussing world energy issues is the quad, short for quadrillion. A quad is equivalent to 10^{15} (1,000,000,000,000,000) Btu, or 1.055×10^{18} (1,055,000,000,000,000,000) J. As a point of reference one quad of energy is produced by the burning of 40,000,000 tons of coal or 27,800,000 tons of oil.
3. Because a discussion of energy issues often involves very large numbers, the use of the powers of ten notation is common in the literature. The powers of 10 notation is a method for expressing very large (and very small) numbers is a simple kind of shorthand. Consider the number one million: 1,000,000,000. The powers of 10 notation expresses that

number in two parts, a whole or decimal number and some power of 10. A power of 10 is the base number 10 raised to some exponent, such as 10^1, 10^2, 10^3, or 10^9. The exponent indicates the number of times 10 is multiplied by itself, so that $10^1 = 10$; $10^2 = 10 \times 10 = 100$; $10^3 = 10 \times 10 \times 10 = 1{,}000$; and $10^9 = 10 \times 10 \times 10 \times 10 \times 10 \times 10 \times 10 \times 10 \times 10 = 1{,}000{,}000{,}000$. You might notice that the exponent is also equal to the number of zeroes in the expanded number: 10^9 = the number 1 followed by nine zeroes.

4. Mass in energy discussions is often expressed in *tonnes,* or *metric tons.* An approximation conversion factor useful in converting tonnes to short tons in the British system is 1 tonne = 1.1 short tons.

The Age of Fossil Fuels: Coal

The segment of human history that begins with the start of the Industrial Revolution is sometimes called the Age of Fossil Fuels because so many human activities depend so completely on the use of coal, oil, and natural gas. The history of these fuels during the 20th century illustrates this point. At first, coal continued to be the dominant energy resource throughout the developed world (although not necessarily among developing nations, which still relied largely on biomass as a fuel source). In Great Britain, for example, the production and the consumption of coal and coal products continued to grow until the middle of World War I, at which point (1913), they began to decline. In that year, British coal companies produced the largest amount of coal in history, 260 million tons, of which the greatest amount (36.9 million tons) went to the production of iron and steel and the next largest amount to ship transportation (23.4 million tons). Other coal consumers were the coke industry (19.7 million tons), mines (18.0 million tons), gasworks (17.8 million tons), railways (14.3 million tons), and electricity generation (4.9 million tons). By 2009, that total had dropped to just 19.2 million tons ("King Coal" 2012).

The United Kingdom's experience in coal production is by no means typical of that of other nations. In the United States, for example, coal production peaked at a 50-year high (starting in 1890) at 677 million tons, after which it fell to its lowest point in 1932 of 359 million tons. The onset of World War II once again created a demand for the fuel, however, and production numbers began climbing again in 1939. After remaining relatively constant after the end of the war at about 400–500 million tons annually, coal production again began to soar in the 1980s until it reached its current high (2010) of 1,085 million tons. Most experts now believe that America's reliance on coal will continue into the foreseeable future (Averitt 1969, 60).

Other nations have experienced coal production patterns that are different from those of the United Kingdom and United States. The story in India, for example, begins with a minuscule production of 6.1 million tons of coal in 1900, rising slowly, but consistently, over the century to 80.9 million tons in 1972 and to 216.1 million tons in 2010, making it the third largest coal producer in the world after China and the United States. This pattern is reflected in even more dramatic ways in two other developing nations, China and Indonesia. The former nation had essentially no coal industry to speak of prior to 1950, with production amounting to only a few tens of millions of tons. That number rose consistently and dramatically over the next half century until China became the world's largest coal-producing nation in 2010, with an output of 1800.4 million tons. The situation has been even more dramatic in Indonesia, whose coal output as recently as 1983 was essentially zero. That nation has now become fourth largest producer, with a 2010 output of 188.1 million tons (Coal Production 2012).

Types of Coal

Like all kinds of fossil fuel, coal is formed when organic matter (dead plants and animals) decays in the absence of oxygen.

Normally, when an organism decays above ground, where it is exposed to air, its organic compounds are converted primarily into carbon dioxide and water, which escape into the atmosphere. In the absence of air, however, those chemical reactions take place to only a limited extent, if at all. Instead, the organic matter is converted into pure carbon and hydrocarbons, compounds that consist of carbon and hydrogen. Methane, ethane, ethylene, acetylene, benzene, and anthracene are only a few examples of the hundreds of thousands of organic compounds produced during the decay of organic materials in the absence of oxygen.

Various types of coal are distinguished from each other based on how far along the process of decay has advanced in a product. The simplest form of coal is peat, often found in damp, boggy areas, where plants materials have died, sunk below the water, and just begun to decay. In fact, peat is often defined not as a form of coal at all, but as pre-coal. The next most advanced form of coal is lignite, also called brown coal. Lignite has a lower proportion of hydrocarbons and a higher (although very low) proportion of pure carbon than peat. Continuing up the scale, with forms of coal that contain increasingly larger fractions of pure carbon and smaller fractions of hydrocarbons, are subbituminous coal, bituminous (soft) coal, and anthracite (hard) coal. The carbon content of bituminous coal ranges from about 60 to about 80 percent, whereas anthracite consists of about 92 to 98 percent pure carbon. The greater the carbon content of a coal, the more heat it produces when burned and, therefore, the more desirable it is for most industrial and other uses.

Mining Technology

The history of coal mining technology is essentially the same as it is for the history of any mining technology. At first, miners extract as much of a resource, such as coal, as is available at or near the ground level. For example, in some parts of the world at one time in history, deposits of coal could be found exposed at ground level. A person could simply pick up a quantity of coal and carry it away. Reserves of coal (and all other resources) of this type disappeared long ago. Over time, miners found that they had to dig deeper and deeper into the ground to find and extract coal, or any other resource. Two early types of coal mines were bell mines and drift mines. In bell mining, a worker simply digs downward into the earth until he or she reaches a vein of coal and then starts removing the coal. Over time, the pit becomes deeper with a wider opening at the top and a narrower diameter at the bottom, its bell shape accounting for the name of the technology. Drift mining is a method of resource removal in which workers dig horizontally into a coal vein, often beginning at the side of a hill. Both bell and drift mining are relatively safe, simple methods of extracting coal.

Again, virtually all coal available near ground level was excavated decades or centuries ago. Eventually, miners had to dig deeper and deeper into the earth to reach a coal vein. The deepest coal mine ever dug, now closed, is reputed to have been a mine in Springhill, Nova Scotia, at a depth of more than 6,500 feet. The mine was closed in 1958 because it was thought to be too dangerous to continue digging. The deepest coal mine in the United States currently is a mine near Brookwood, Alabama, about 2,100 feet below ground (about the height of a 175-story building).

A variety of technologies are available for removing coal from underground sites. Perhaps the best-known and the most common is called the room-and-pillar technology. In this method of mining, workers dig downward until they reach a coal vein and then begin to cut out sections of coal and move it to the surface. As larger sections of the vein are removed, the site begins to look like a large conference room. The process of extraction is halted at certain points to leave large posts of coal to support the overburden of earth above the vein of coal. As mining progresses, the vein begins to look like a number of large rooms that are supported by pillars of coal. Those pillars can never be removed as long as mining continues because they prevent the overburden from collapsing into the emptied vein.

Underground mines are associated with a number of major issues. One, of course, is the cost of digging deep holes in the earth and extracting the resource. A second problem is the environmental damage that can result when the pillars in a room-and-pillar mine collapse, causing the mine itself to collapse. In some cases, this type of *subsidence* can cause extensive damage to homes and other buildings on the surface above the mine. By far the most serious problem, however, is the risk to workers in an underground mine. The history of underground coal mining is studded with an unending series of disasters resulting from the collapse of a mine ceiling or fire in an underground mine. The most deadly coal mining disaster in history occurred at the Benxihu (Honkeiko) Colliery in Benxi, Liaoning, China on April 26, 1942. A coal dust explosion in the mine resulted in the death of 1,549 workers, about one-third of the crew working that day. More than 30 other coal mine disasters worldwide have resulted in more than 200 deaths each, and more than 60 have resulted in the death of more than 100 miners each. The worst coal mining disaster in U.S. history occurred at the Monongah (West Virginia) Mine on December 6, 1907, an accident in which 362 miners were killed. The accident is thought to have been caused by the explosion of firedamp (methane gas). By some estimates, more than 100,000 workers were killed in coal mine accidents during the 20th century (Ngô and Natowitz 2009, 87).

In an effort to minimize both costs and mining accidents (as well as for other reasons), companies have shifted over time from subsurface (underground) to surface mining. Four types of surface mining are common: open pit, mountain-top removal, strip, and auger. Open pit mining is a modified form of traditional bell pit mining in which the bell shape is expanded to much greater dimensions. Overburden is first removed and transferred to some location outside the actual mining site. Coal is then removed by digging it out of the vein at deeper and deeper depths and wider and wider horizontal distances, producing the classic bowl-shaped appearance of the open pit mine. The largest open pit coal mine in the world is located in Cerrejón, Colombia. The facility employees nearly 10,000 workers in direct operations, with an additional 30,000 to 50,000 employees in support activities. The pit produced 33.3 million tons of coal in 2010 and has an estimated total reserve of 2,161 million tons of coal.

A second type of surface coal mining is mountain-top removal, which, as its name suggest, involves cutting away a substantial portion of a mountain top in order to get at coal seams within the mountain. The process begins by scraping away as much as 1,000 feet of overburden (the mountain top) and transferring that material to an adjacent valley. The coal thus exposed can then be removed directly by large excavating equipment. A third type of surface mining is called strip mining. As with mountain-top removal mining, the first step is to strip away the overburden and then begin removing the coal beneath that overburden, except that the process occurs on flat or nearly flat land. When the stripping occurs in a horizontal direction, the process is called area stripping, and when it is carried out along the slope of a hill or a mountain, the process is called contour stripping. A fourth type of surface mining, auger mining, involves cutting into the end of a coal seam that is exposed at the side of a hill or a mountain or that has already been mined by some other technique. The auger used in auger mining is a very large drill that cuts away coal from the seam and then carries that coal away on a moving track.

Legislation

A comprehensive review of coal mining laws around the world is beyond the scope of this book. Suffice to say that those laws range from the limited and weakly enforced to the extended and strongly enforced in various parts of the world. Laws in the United States and Western Europe tend to be the most comprehensive and detailed of any in the world. In the United States, for example, coal mining is covered to some extent by more than 50 major federal laws and executive orders ranging from the American Indian Religious Freedom Act to the Bald and Golden Eagle Protection Act to the Noise Control Act to the Noxious Weed Act. Among the most important coal mining legislation deals with mine safety. The first U.S. legislation on this subject was passed by the U.S. Congress in 1891. It established standards for mine ventilation and prohibited the employment of children under the age of 12 in U.S. mines. Far more aggressive legislation was adopted in 1910 after a decade in which mine deaths mounted to an average of about 2,000 per year. The 1910 act established the Bureau of Mines, charged with finding ways of reducing coal mining accidents. The bureau was hampered to some extent because legislation permitting mine inspection was not adopted until more than three decades later, in 1941. The birth of even modestly effective coal mining legislation in the United States dates to 1952 with the passage of the Federal Coal Mine Safety Act, which gave the Bureau of Mines authority to inspect some (but not all) coal mines on an annual basis, along with limited authority to issue notices of violation and assess fines for those violations. The law was extended in its scope and intensity in the Federal Coal Mine Health and Safety Act of 1969, which brought surface mining under federal inspection along with subsurface mines. In 1973, the Secretary of the Interior established the Mining Enforcement and Safety Administration (MESA) to separate two conflicting missions of the Bureau of Mines: health and safety, on the one hand, and the promotion of mining activities, on the other. MESA was later to become the Mine Safety and Health Administration (MSHA), whose mission was established in the Federal Mine Safety and Health Administration Act of 1977. That act provides the general framework for the nation's current policies and practices in coal mine safety enforcement activities. The act was amended and strengthened in the Mine Improvement and New Emergency Response Act (MINER Act) of 2006 ("History of Mine Safety and Health Legislation" 2012).

Current Status of the Coal Industry

Three categories are of special interest in describing the current status of the worldwide coal industry: reserves, production, and consumption. The term *reserves* is widely used and very important in the energy industry; the term also has a variety of meanings that range across the spectrum depending on how certain are that those reserves can also be harvested to produce coal, oil, or natural gas. For example, reserves are sometimes classifieds *proved* (or *proven*) reserves and *probable* reserves. The former term expresses a higher degree of confidence that the resource (coal in this case) can be extracted from the reserve with technology currently available. As of 2011, the largest proved reserves of coal in the world are those found in the United States, with an estimated 119,602 million tons of anthracite and bituminous coal and an estimated 140,869 million tons of subbituminous and lignite. These numbers account for 27.6 percent of all the proven coal reserves in the world. The next nations with the next largest proven reserves of coal are the Russian Federation (18.2% of the world total), China (13.3%), Australia (8.9%), India (7.0%), Germany (4.7%), and Kazakhstan and Ukraine (3.9% each) (*BP Statistical Review of World Energy* 2012, 30).

The largest coal producing nation in the world in 2010, by far, was China, with a total output amounting to 1,984.6 million tons of oil equivalent (a unit sometimes used to express coal production). This number accounted for almost half (48.3%) of all the coal produced in the world in 2010. Other major coal-producing nations after China were the United States (608.7 million tons; 14.8% of world total), Australia (259.5 million tons; 6.3%), India (238.2 million tons; 5.8%), and Indonesia (207.3 million tonnes; 5.0%) (*BP Statistical Review of World Energy* 2012, 32).

Some similarities exist between coal-producing and coal-consuming nations in 2010. Again, China was by far the largest consumer of coal in the world, using 1,888.8 million tons, or 48.2 percent of the world total. It was followed by the United

States (578.3 million tons; 14.8% of world total), India (306.0 million tons), Japan (136.4 million tons), Russian Federation (103.4 million tons; 2.6%), and South Africa (97.8 million tons; 2.5%) (*BP Statistical Review of World Energy* 2012, 33).

A final measure of some interest to analysts is the so-called *reserve-to-production* (or *R/P*) *ratio.* This ratio compares the size of a nation's proven reserves of coal (the *R* in the fraction) to its current rate of production (the *P* in the fraction) annually. The ratio is often used as a predictor of the period of time over which some resource (such as coal) will be available to a country. For example, suppose that a country has proven reserve of 100 units of coal and is using 10 units of coal annually. The *R/P* ratio for that country is 100/10, which means that the country is expected to run out of that resource in 10 years. Of course, both the *R* and the *P* in this ratio continue to change from year to year as more reserves are found (increasing the value of *R*) and the rate of consumption increases or decreases. For 2010, the nations with low *R/P* ratios who are presumably at risk for running out of coal were nations such as Vietnam (*R/P* = 3), Romania (*R/P* = 9), United Kingdom (*R/P* = 13), North Korea (*R/P* = 16), and Indonesia (*R/P* = 18). An important nation at this end of the scale was China, with an *R/P* = 35. At the other end of the scale were nations whose resources, either large or small, were expected to last much longer periods of time, such as the Russian Federation (*R/P* = 495), Ukraine (*R/P* = 462), Japan (*R/P* = 382), and Kazakhstan (*R/P* = 303). The *R/P* ratio for the United States in 2010 was 241, suggesting that the nation's coal reserves at the present time should last about 241 years (*BP Statistical Review of World Energy* 2012, 30)

The Age of Fossil Fuels: Petroleum

While the 19th century was clearly the age of coal, the 20th century can best be described as the age of petroleum. By the end of the 19th century, there were relatively few commercial uses for petroleum products, one of the most important of which was the use of kerosene for lighting. Even that market began to disappear with the invention of the electric light bulb by Thomas Alva Edison in 1882. It may have seemed at the time that drilling for oil was a waste of money. Within a decade, however, that type of prediction was proved wrong by the invention of the internal combustion engine, which made possible the manufacture of the modern motor vehicle. The first great commercial success in this regard was the production of the first motor car by Henry Ford in 1896. Suddenly, a new use was found for petroleum, the production of gasoline for the operation of motor vehicles.

The contribution of the motor car to the development of the petroleum industry was not at first clear. Of the 4,192 cars manufactured in the United States in 1900, only about a quarter (936) were gasoline-driven vehicles. The others were either steam-driven (1,681; 40.1%) or electric cars (1,575; 37.6%). The United States was not the only country where motor vehicle production was underway. In fact, until 1903, the largest motor-vehicle industry was in France, which turned out 30,204 cars in that year, 48.8 percent of the total world production for the year. Other countries with significant car-production facilities at the time were Australia, Belgium, Denmark, Italy, Norway, Sweden, and Switzerland. These companies were turning out motor vehicles not only for domestic sale, but also for export to countries around the world, including the Dutch East Indies (now Indonesia), Egypt, Iran, and Tunisia ("History of Motor Car/Automobile Production 1900–2003" 2012).

The automobile industry experienced dramatic successes in the first quarter of the 20th century with perhaps as many as 1,800 companies founded during the period for the production of cars and trucks. At least those numbers were being formed worldwide to meet the growing demands for this new form of transportation. Only a small fraction of those numbers survived for more than a few years. Still, the industry continued to grow throughout Europe and the United States. By 1910, more than 100,000 cars were registered in Great

Britain, 47,000 in Germany, and 495,000 in the United States. The U.S. number more than tripled to 1.7 million vehicles in 1914, at the beginning of World War I.

The onset of the war in 1914 provided another enormous impetus for the development of the petroleum industry as gasoline was required for most of the new forms of transportation developed for military use, such as cars, trucks, railroad engines, and airplanes. These vehicles were much preferred over the traditional method for moving men and goods—the horse. At the war's beginning, however, Great Britain had only 800 motor vehicles available for military use, nearly all of them requisitioned from private owners. During the war, however, the British built and put into service and additional 36,000 cars and 56,000 trucks, numbers that were nearly matched or exceeded by other nations involved in the conflict. At the same time, the United States contributed its own share of gasoline-based vehicles—over 50,000 cars and trucks and 15,000 airplanes—in the 18 months during which it was involved in the war. This sudden flood of motorized vehicles of course produced a corresponding demand for the fuels needed to drive and lubricate them, a boon to the petroleum industry. During the war, the various elements that made up that industry began cooperating with each other and with the governments of the nations involved to bring about victory for one or the other side in the conflict. One of the great economic consequences of the war, then, was a significant boost to the young petroleum industry.

The Petroleum Industry

At the onset of World War I, the petroleum industry was already a half century old, although it was hardly the robust economic engine that it became in the 21st century. The world's first oil company was the Pennsylvania Rock Oil Company of New York, founded in 1854 by New York lawyer George Bissell. Bissell had hired the famous Yale University scientist Benjamin Silliman Jr. to analyze a sample of oil taken from a field in Titusville, Pennsylvania. Silliman's analysis convinced

Bissell and a group of bankers of the potential value of the field as a source of oil, prompting them to establish the company. It was under the auspices of Pennsylvania Rock Oil that Edwin Drake produced the first oil well in the United States in 1859. Pennsylvania Rock Oil was later reconstituted as the Seneca Oil Company, which survived until 1986, when it was acquired by the Kaiser-Francis Oil Company of Tulsa, Oklahoma.

The period between 1866 and 1897 was the heyday for the creation of a number of petroleum companies in the United States and other parts of the world. Many of the petroleum giants of the 21st century trace their origins to that period. By some standards, the oldest of those companies is modern-day Shell, whose earliest history dates to the creation in 1833 of a small shop in London by entrepreneur Marcus Samuel for the sale of seashells. On a visit to the Caspian Sea in search of shells for his father's shop, Samuel's son, also named Marcus, realized the potential for exporting oil to the Far East for cooking and lighting. The seashell shop eventually evolved into a more ambitious company for the distribution of petroleum products, and the Shell Transport and Trading Company, Ltd., was founded for that purpose in 1897. A decade later, Shell merged with its primary competitor in the Far East market, Royal Dutch, which had been founded in 1890. A century later, after a number of mergers and acquisitions, the company remains one of the largest petroleum companies in the world with its world headquarters in Houston, Texas.

The following lists other petroleum companies formed during the 1860s, 1870s, 1880s, and 1890s.

- In 1866, the Atlantic Petroleum Storage Company, which later became part of Standard Oil, was reconstituted as the Richfield Oil Corporation, acquired the Sinclair Oil Corporation, and eventually became the Atlantic Richfield Company, today's ARCO.
- Also in 1866, the Vacuum Oil Company, which was absorbed by and then dissociated from Standard Oil, acquired a number of other smaller companies, and was eventually

acquired itself by the Standard Oil Company of New York (Socony), which later became Socony Mobil, and later Mobil Oil Company. As a result of a later merger with Exxon, the company, now known as Exxon Mobil, had the largest revenue of any publicly traded company in the world with a net income in 2010 of $7.56 billion.

- In 1870, the Standard Oil Company was formed, about which more can be found subsequently.
- In 1875, the Continental Oil and Transport Company was formed, later absorbed by and dissociated from Standard Oil, and, after various reorganizations, merged with the Phillips Petroleum Company in 2002 to form ConocoPhillips, the 12th largest publicly traded company in the world.
- In 1879, the Pacific Oil Company was established, later to become a part of Standard Oil before reincorporating as Standard Oil of California (Socal), and buying Gulf Oil in 1984 and becoming the Chevron Corporation, the 11th largest publicly traded corporation in the world today.
- In 1882, Standard Oil of New Jersey was formed, which, after a number of acquisitions and mergers, became the Exxon corporation in 1972.
- In 1886, the Burmah Oil company is formed in Scotland, after which its name is changed first to the Anglo-Persian oil company, then the Anglo-Iranian oil company, then, in 1954, British Petroleum (BP). In 1998, BP merged with Amoco to become BP Amoco (currently known just as BP), now the fourth largest publicly traded corporation in the world with revenue in 2010 of $297 billion.

This list is by no means complete, although it does provide a skeletal history of some of the largest oil corporations in existence today. One company stands out from that list and requires additional comment: the Standard Oil Company in its various configurations and manifestations. The early history of Standard Oil is, to a large extent, the history of John D. Rockefeller. Rockefeller first became involved in the oil business in 1863 when he and a group of partners built an oil refinery in an area of Cleveland, Ohio, known as The Flats. Their plan was to deliver refined oil to the East Coast by way of the Atlantic and Great Western Railroad. In 1865, Rockefeller bought out his partners and created Rockefeller & Andrews, at the time, Cleveland's largest refinery. In 1868, Rockefeller made a sweetheart deal with Jay Gould, owner of the Erie Railroad, by which he was able to ship his oil to the East Coast at significantly reduced rates in return for breaks on fuel sold to the railroad. Two years later, Rockefeller founded the Standard Oil Company, at which point the company was refining about 10 percent of all the oil in the United States.

Between 1872 and 1873, Rockefeller made a number of deals that allowed him to buy up virtually all of the refineries in Cleveland, Pittsburgh, Philadelphia, New York, and other parts of the east. By 1877, Rockefeller's refineries were producing 90 percent of the refined oil in the United States. In 1882, the bulk of Standard's many and far-flung operations were brought together in the Standard Oil Trust, whose functioning was not well understood by regulators or the general public. The company continued to grow over the next three decades, even after Rockefeller himself retired from active participation in the business in 1895.

In 1907, the federal government finally decided to take action against the Standard Oil Trust because it had cornered such a large share of the market. Its critics complained that, at the time, the corporation was 20 times larger than its next nearest competitor. Acting under the provisions of the Sherman Anti-Trust Act, the government filed seven different suits against the trust, demanding that it be dismantled. Four years later, the U.S. Supreme Court finally resolved this legal question, requiring that the trust be dissolved and that it divest itself of all of its subsidiaries. The break-up of the trust resulted in the formation of seven new companies: Standard Oil of New Jersey, Standard Oil of New York, Standard Oil of California, Standard Oil of Ohio (later acquired by British Petroleum),

Standard Oil of Indiana (later to become Amoco), Continental Oil (later, Conoco), and Atlantic Oil (later, Arco) ("Whatever Happened to Standard Oil?" 2012).

This discussion gives a somewhat distorted view of the petroleum business worldwide since it includes only privately held companies, such as Exxon Mobil and BP. In addition to these companies are a number of state-owned oil companies that began their lives as subsidiaries of the large publicly owned companies. As an example, the largest of the state-owned companies, Saudi Aramco, began life in 1933 when the Saudi Arabian government granted oil concessions to Standard Oil of California, which then created the wholly owned subsidiary Californian Arabian Standard Oil Company (Casoc). Casoc later sold half of its interests in the project to the Texas Oil Company (Texaco) in 1936. The company changed its name to the Arabian American Oil Company (or Aramco) in 1944.

As Arab states began to develop and mature, they became more interested in having greater control over their own fossil fuel resources and began to nationalize the previously privately owned oil companies. Saudi Arabia purchased a 25 percent share of Aramco in 1973, following the Yom Kippur war between Israel and its Arab nations. It increased its share to 60 percent in 1974 and finally took complete control of the company in 1980. Today, that company, now known as Saudi Aramco, is thought to be the largest corporation in the world with assets estimated at between $2.2 and $7 trillion. The evolution of oil companies from privately owned, publicly traded entities (such as Standard Oil of California and Texaco) to state-owned monopolies has followed the same course in virtually all other oil-producing nations.

Production and Consumption Trends

The Age of Petroleum can perhaps be said to have begun in the 1860s with the recent discovery of oil in the Baku region of Azerbaijan and, primarily, in Pennsylvania in the United States.

For most of the next century, the United States and Azerbaijan (and, after it was absorbed, by Russia) dominated world production of petroleum. The United States was the lead producer during most of that period, although Azerbaijan (Russia) took that honor in a few years.

Oil production in the early part of the period was minuscule, averaging about less than 400 bpd (barrels per day) in 1860. It grew very slowly at first, reaching 10,000 bpd in 1866 and 100,000 bpd in 1885. By 1900, worldwide production of petroleum exceeded 410,000 bpd. (For comparison, worldwide petroleum production at the end of 2011 was just over 89.3 million barrels per day.) During and following World War I, petroleum production began to accelerate worldwide reaching 1 Mmbpd (million barrels of oil per day) in 1913, 2 Mmbpd in 1921, and 6 Mmbpd in 1941 (Yergin 1991, part I).

For the first two decades of the 20th century, the third leading producer of petroleum in the world was Mexico, where exploratory wells were dug as early as 1869. By 1901, some of these wells were productive, and the country became a major oil producer. Mexico's oil production continued to grow throughout the 20th century and reached its peak level in 2004, when production reached 3,824 Mmbpd. It has since declined by a cumulative 2.9 percent to its 2010 level of 2,953 Mmbpd. In 1938, the Mexican government nationalized oil production in the country, placing all petroleum operations under the authority of the agency Petróleos Mexicanos (Pemex), which continues to serve as the national petroleum agency today (Gonzalez 2012).

In the period between 1921 and 1940, Mexico fell to fourth place in world oil production, replaced by Venezuela. The first oil wells were dug in Venezuela at Lake Maracaibo in 1878. Again, the industry grew slowly until it reached less than a half million barrels per day in 1930, 1 Mmbpd in 1950, 2 Mmbpd in 1955, and 3 Mmbpd in 1960. Venezuela was one of the founding members of Organization of Petroleum Exporting Countries (OPEC), and the oil industry was nationalized in 1976 under President Carlos Andrés Pérez. Petróleos de Venezuela S.A. (PDVSA) was created to oversee the industry, which it continues to do in 2012 ("History of Venezuela's Oil and Rentier Economy" 2012).

The Middle Eastern states, currently the dominant force in world petroleum production, came somewhat late to the industry. Oil was first discovered in Iran in 1908, in Iraq in 1927, in Bahrain in 1932, in Kuwait and Saudi Arabia in 1938, and in other now-major oil-producing states, such as Algeria, Nigeria, and the United Arab Emirates, not until the 1950s. In the period between 1941 and 1960, some of these states began to move to the top of the list of world oil-producing states. During that period, Saudi Arabia took fourth place, after the United States, Venezuela, and the Soviet Union, followed by Iran in fifth place. Table 1.2 shows the relative ranking of the top five oil-producing countries in the world in the second half of the 20th century and the first decade of the 21st century. (Rankings for any specific year within a period may vary; these rankings show two-decade trends.)

As the table shows, the world's largest producer of petroleum products in 2011 was Russia with an output of 10.1 Mmbpd, followed by Saudi Arabia (9.76 Mmbpd), the United States (9.06 Mmbpd), Iran (4.17 Mmbpd), and newcomer China (3.99 Mmbpd).

The United States has always been the world's largest consumer of oil. As far back as 1965, for example, the United States was using 11.5 Mmbpd of oil, with the next largest consumers

Table 1.2 Top Five Oil-Producing Countries, 1941–2011

1941–1960	1961–1980	1981–2000	2001–2010	2011
United States	United States	Soviet Union	Saudi Arabia	Russia
Venezuela	Soviet Union	United States	Russia	Saudi Arabia
Soviet Union	Saudi Arabia	Saudi Arabia	United States	United States
Saudi Arabia	Iran	Iran	Iran	Iran
Iran	Venezuela	Mexico	Mexico	China

Source: Vassilious (2009, 6–7).

Table 1.3 Top Oil-Consuming Countries, 1980–2010 (in Mmbpd)

1970	1980	1990	2000	2010
United States (14.71)	United States (17.06)	United States (16.99)	United States (19.70)	United States (19.15)
Japan (3.87)	Japan (4.90)	Japan (5.23)	Japan (5.53)	China (9.06)
Germany (2.77)	Germany (3.02)	Russian Federation (5.05)	China (4.77)	Japan (4.45)
United Kingdom (2.03)	France (2.22)	Germany (2.69)	Germany (2.75)	India (3.32)
France (1.87)	Italy (1.93)	China (2.32)	Russian Federation (2.70)	Russian Federation (3.20)

Source: BP Statistical Review of World Energy (2012).

being Japan (1.7 Mmbpd), Germany (1.7 Mmbpd), and the United Kingdom (1.1 Mmbpd). As Table 1.3 shows that pattern has changed relatively little over the years.

The data in Table 1.3 represent an interesting shift in the consumption of oil worldwide. Prior to the 21st century, the leading consumers of oil outside the United States and Japan were virtually all in Western Europe. As detailed 2010 data show, however, more of the nations once thought of as developing nations are now large oil consumers, including not only those nations listed in Table 1.3, but also Brazil (2.60 Mmbpd in 2010), Iran (1.80 Mmbpd), Saudi Arabia (2.81 Mmbpd), and South Korea (2.38 Mmbpd), all with consumption totals greater than those of any Western European country except Germany (2.44 Mmbpd in 2010).

Petroleum Technology

Like all fossil fuels, petroleum was formed when organisms died and were buried under water or underground, where they decayed in the absence of air (oxygen). Unlike coal, this decay generally resulted in the form of gaseous or liquid hydrocarbons (natural gas or petroleum) rather than pure solid carbon. These liquid and gaseous hydrocarbons were trapped within underground rock layers and eventually migrated to the most porous of those rocky materials, such as sandstone and limestone. Over millennia, rock movements tended to concentrate these materials into hydrocarbon-rich regions, from which they can be extracted today. That extraction process consists essentially of two steps: locating the oil and gas and extracting it from the earth.

In the early history of human civilization, finding oil was often not a problem: the material seeped to the surface and formed small pools from which the resource could be removed directly. In other instances, it took little more than poking a stick into the ground (a la "The Beverly Hillbillies") to gain access to a source of petroleum. Petroleum that was that readily available was exhausted very early in human history, and the earliest oil wells in the 19th century were dug, at first, with shovels, or, very soon thereafter, with some type of drilling devices. Most 19th-century oil-well drillers used a technology that had been developed many decades earlier for the production of water wells. A cable tipped with a touch carbide drill is dropped into a hole and then set to drilling by some type of power, such as in the case of Edwin Drake's first oil well, a steam engine. Over the years, a number of technological improvements have been made in this model, and oil wells can now be drilled to depths of almost seven miles. The deepest hole ever recorded is one drilled by the offshore drilling rig *Deepwater Horizon* in 2009, which eventually reached a depth of 35,055 feet (10,683 meters). (It was that drilling rig that caught fire and burned with devastating environmental consequences in April 2010.)

Although the early history of oil drilling is that of on-land drilling, the prospect of collecting oil from beneath bodies of water has been obvious to explorers for centuries. Probably the first of these offshore wells was drilled in the Caspian Sea, about 100 feet from shore, in 1803. The wells produced oil from shallow wells until they were destroyed by a storm on the sea in 1825. A number of other attempts at offshore drilling were made during the rest of the 19th century and into the early 20th century, including wells drilled from platforms on Grand Lake St. Marys in Ohio in 1891, from the end of piers in the Santa Barbara Channel in California in 1896, and on the Canadian side of Lake Erie in the 1900s and Caddo Lake in Louisiana in the 1910s. By the 1940s, engineers had designed and built the earliest models of large offshore drilling rigs with which most people are familiar today, capable of drilling hundreds or thousands of meters into the ocean bottom tens or hundreds of miles offshore. As of February 2012, there were 1,989 drilling rigs in U.S. territories, 709 in Canada, and 1,171 rigs elsewhere in the world ("Overview & FAQ" 2012).

The Age of Fossil Fuels: Natural Gas

Natural gas is a type of fossil fuel formed like coal during the anaerobic (without oxygen) decomposition of organic matter. It is almost always found in combination with petroleum and coal. Natural gas consists primarily of hydrocarbons, compounds consisting of carbon and hydrogen only, such as methane (typically about 95% of natural gas), ethane (about 2%), propane, *n*-butane and isobutane, *n*-pentane and isopentane, and isomers of hexane (all constituting less than 1% of a sample of natural gas), with very small amounts of nitrogen, carbon dioxide, oxygen, and other inorganic gases.

As mentioned previously, natural gas has been known to and used by humans for millennia, usually for the lighting of lamps and for ceremonial purposes. The first commercial use of natural gas is usually associated with efforts by American inventor and entrepreneur William A. Hart who, in 1821, constructed a simple system for providing the city of Fredonia, New York, with a source of natural gas for lighting its homes and businesses. Hart drilled a 27-foot hole into a naturally occurring source of natural gas and constructed a system of pipes to deliver the gas to buildings in the community. An important early development in the commercialization of natural gas as a fuel was the invention of the so-called Bunsen burner, a device that mixes natural gas with oxygen to produce a flame that burns more cleanly and hotter than does the combustion of natural gas in air. (The name of the device is somewhat of a misnomer since it was not actually invented by the great German chemist Robert Bunsen himself, but by one of his assistants, Peter Desaga. Nonetheless, the burner will undoubtedly forever be known by its present name.) ("History" 2012.)

The commercialization of natural gas as a source of energy for lighting, heating, cooking, and other purposes was delayed for a relatively long time, primarily because of the problems involved in transporting gas from source to destination. Hart had originally used wooden pipes joined with pitch, but gas still leaked through the system rather easily. This problem was not really solved adequately until 1891 when inventor and industrialist Elwood P. Haynes oversaw the construction of a 120-mile-long (190-kilometer-long) pipeline to carry natural gas from the Trenton Gas Field in Indiana to the city of Chicago (Gray 1995, 208–14). This impressive accomplishment stood essentially alone, however, as the development of the electric light bulb by Thomas A. Edison in 1879 provided a new option for lighting, which greatly reduced the demand for natural gas. As a consequence, pipeline construction developed relatively slowly over the next century. A report issued in 1894, for example, said that 27,350 miles of pipeline were then being used to transmit natural gas from fields to consumers. In 1912, one of the longest gas pipelines to date was built to transport natural gas from the Bow Island gas field in Alberta first to Lethbridge, then on to Calgary, a distance of 160 miles (270 kilometers). In 1925, the first all-metal welded gas pipeline between Louisiana and Beaumont, Texas, a distance of 217 miles (347 kilometers) was opened. It was not until the end of World War II, however, that technology had advanced to the point where reliable, long-distance pipelines could be built and with that technology, pipelines began to spring up rapidly around the world. As of early 2012, the longest natural gas pipeline in the world is one that connects Shanghai, China, with a gas field in Xinjiang Autonomous Region, a distance of 5,410 miles (8,660 kilometers). The longest gas pipeline in the United States runs from Colorado to Ohio, a distance of 1,678 miles (2,685 kilometers), with plans to extend the pipeline to the East Coast, roughly doubling its length ("Natural Gas: The Golden Age of Gas?" 2012).

Production and Consumption Trends

At the end of 2010, the world's largest producer of natural gas was the United States, with production of 21,580 bcf (billion cubic feet) or 19.3 percent of the world total for that year. The next largest producers were the Russian Federation (20,780 bcf; 18.4%), Canada (5,643 bcf; 5.0%), Iran (4,901 bcf; 4.3%), Qatar (4,121 bcf; 3.6%), and Norway (3,758 bcf; 3.3%). Table 1.4 shows historical trends in the top five natural gas producing countries in the world since 1970.

Patterns for the consumption of natural gas, as might be expected, tend to follow those for petroleum. Table 1.5 shows changes in the top five consuming nations for natural gas between 1970 and 2010.

An interesting trend toward the end of the 20th century in natural gas use, which mirrors that for petroleum use, does not show up in Table 1.5. This trend is the moderation in the

Table 1.4 Top Natural Gas-Producing Countries, 1970–2000 (in bcm)

1970	1980	1990	2000	2010
United States (595.1)	United States (549.4)	Russian Federation (590.0)	United States (543.2)	United States (611.0)
Canada (56.7)	Netherlands (76.4)	United States (504.3)	Russian Federation (528.5)	Russian Federation (588.9)
Netherlands (26.7)	Canada (74.8)	Canada (108.6)	Canada (182.2)	Canada (159.8)
Romania (23.3)	United Kingdom (34.8)	Turkmenistan (79.5)	United Kingdom (108.4)	Iran (138.5)
Iran (12.9)	Romania (34.7)	Netherlands (61.0)	Algeria (84.4)	Qatar (116.7)

Source: BP Statistical Review of World Energy (2012).

Table 1.5 Increase in Natural Gas Consumption for Certain Nations, 1970–2010 (bcm)

1970	1980	1990	2000	2010
United States (598.6)	United States (562.9)	United States (542.9)	United States (660.7)	United States (683.4)
Canada (36.4)	Germany (57.4)	Russian Federation (407.6)	Russian Federation (354.0)	Russian Federation (414.1)
Romania (23.2)	Canada (52.2)	United Kingdom (124.0)	United Kingdom (96.9)	Iran (136.9)
Netherlands (16.9)	United Kingdom (44.8)	Canada (67.2)	Canada (92.7)	China (109.0)
Germany (15.0)	Romania (36.0)	Germany (59.9)	Germany (79.5)	Japan (94.5)

Source: BP Statistical Review of World Energy (2012).

growth for the demand for both petroleum and natural gas in the developed nations of Western Europe and Japan, accompanied by an increase in the demand for these fuels in the Far East. This trend is suggested by the annual increase in natural gas use by certain countries in the region, as shown in Table 1.6.

Some experts now argue that natural gas may be the best possible option for solving the world's short-term energy problems. The fuel burns cleanly with relatively few pollutants, and it is

Table 1.6 Increase in Natural Gas Consumption for Certain Nations, 2009–2010 (in bcm)

Country	2009	2010	Percent Increase
United States	646.7	683.4	5.6
Canada	94.4	93.8	–0.6
Germany	78.0	81.3	4.2
Italy	71.5	76.1	6.4
United Kingdom	86.7	93.8	8.3
China	89.5	109.0	21.8
India	51.0	61.9	21.5
South Korea	33.9	42.9	26.5
Taiwan	11.3	14.1	24.3
Vietnam	8.0	9.4	16.7

Source: BP Statistical Review of World Energy (2012).

available abundantly throughout the world. Currently, the largest proved reserves of natural gas are found in the Russian Federation, with an estimated 1,580 trillion cubic feet of natural gas, equal to 23.9 percent of total world reserves of the resource. Other countries with large proved reserves are Iran (1,045 trillion cubic feet; 15.8% of the world total), Qatar (893 trillion cubic feet; 13.5%), Turkmenistan and Saudi Arabia (282 trillion cubic feet; 4.3% each), and the United States (272 trillion cubic feet; 4.1%). Overall, the total proved world reserves of natural gas are estimated to be 6,607 trillion cubic feet, an amount that is expected to last until about 2070 ("Natural Gas: Proved Reserves" 2012). Such estimates are uncertain, of course, because they do not take into consideration the size of future possible discoveries. By comparison, the world's current supply of petroleum is expected to last until about 2058 and the supply of coal, until about 2130.

The technology used in the exploration and extraction of natural gas is similar to that for petroleum, with which it is virtually always found. One important difference between the two fossil fuels is in the method of transportation. Since petroleum is a liquid, it can rather easily be shipped through pipelines, carried by trucks, and stored in tanks. Treating a gas like natural gas in the same way is much more difficult. Natural gas tends to take up a very large volume for its mass and may escape easily from many types of containers. One solution to the transport of natural gas, then, is to first convert it into a liquid, known as liquefied natural gas (LNG). The LNG takes up about 1/600th the volume of the gas itself and can be transported in special trucks, ships, and railroad cars to central distribution sites and then shipped to its final destinations. Gaseous natural gas is converted to LNG by first removing impurities that freeze at the temperatures used to liquefy the gas, such as water, carbon dioxide, and hydrogen sulfide. The purified gas is then cooled to a temperature at which its components (primarily methane and ethane) liquefy. The LNG thus produced can be stored or shipped immediately to its final destinations. Currently, the largest facility in the world for converting natural gas to LNG is the Qatargas II facility in Ras Laffan, Qatar, with a capacity of generating 7.8 million metric tons of LNG annually.

Alternative Fossil Fuels

Coal, petroleum, and natural gas constitute by far the largest portion of the world's fossil fuel supplies. However, these fuels are also available in a number of alternative forms that have the potential for making important contributions to the world's energy equation. Two examples of special interest at the start of the 21st century are shale oil (also sometimes called rock oil) and tar sands. In a somewhat confusing use of terms, shale oil is a liquid fossil fuel obtained from a sedimentary rock known as oil shale. Oil shale contains a solid organic material called *kerogen* formed millions of years ago by the same process by which coal, oil, and natural gas were produced. When oil shale is heated to high temperatures (the process is called *pyrolysis*), solid kerogen changes into a mixture of liquid and gaseous hydrocarbons very similar to the petroleum and natural gas found in an oil deposit. Both the liquid shale oil and the gaseous shale gas can be used directly for a number of purposes, such as power generation or as a heating fuel. Or they can be purified for use in other types of industrial operations, such as fractionation into simpler components or the production of synthetic petroleum products.

Oil shale and shale oil have been known to and used by humans for centuries. Some of their earliest uses were as adhesives in building projects, as caulking for ships, as medicines, in the paving of roads, as fuel in systems for converting salt water to fresh water, and as components in the pyrotechnic material known as "Greek fire." In 1694, the British government awarded a patent to three individuals who had found a way "to extract and make great quantities of pitch, tarr, and oyle out of a sort of stone" (Shale Oil 2012). As whale oil became too expensive in the early 19th century, shale oil was one of the fuels used as its replacement in lighting devices. By the end of the century, most European companies had widespread industrial operations for the production and distribution of shale oil. Many of these operations continued to grow throughout the 20th century, and by 1980, world production of oil shale reached nearly 55 million tons annually. The vast majority of that production occurred in a small number of countries: Estonia, Russia, and China. Interest in the product fell off after 1980 as the price of crude oil dropped, with only these three countries continuing to produce significant amounts of the product.

As the supplies of crude oil and natural gas have declined and will continue to decline, experts have begun to consider the potential of shale oil in replacing some of these traditional energy resources. Certainly vast amounts of oil shale are available throughout the world, by far the largest being in the United States. Four sites in the United States are the largest such reservoirs in the world: the Green River Formation, with estimated reserves of 235,000 million tons; the Phosphoria Formation (39,435 million tons); Eastern Devonian (30,000 million tons); and Heath Formation (28,197 million tons). Other nations with especially large reserves are Russia, the Democratic Republic of Congo, Brazil, Italy, Morocco, Jordan, Estonia, and Canada. Of those reserves, only a few are currently producing shale oil, those in Estonia (308 million gallons in 2008), China (124 million gallons), and Brazil (66 million gallons) (Dyni 2006).

Tar sands are similar to shale oil in that they consist of inorganic materials, such as sand and clay saturated with a thick solid or semi-solid material known as bitumen, or tar. Tar sands are also known as oil sands or bituminous sands. Their physical properties vary somewhat depending on the geographic location of the resource site. In Canada, one of the two nations of the world with the largest reserve of tar sands, the material is so thick that it does not flow unless heated or mixed with lighter liquid hydrocarbons. In Venezuela, the other country with massive reserves of tar sands, temperatures are sufficiently warm to allow the mixture to flow, although only very sluggishly. Because of this property, tar sands are also known as *heavy oil* in Venezuela. According to some estimates Canada and Venezuela each have tar sands reserves equivalent to all of the proven petroleum reserves in the world and, thus, have the potential for playing a critical role in the world's energy equation in the future. Current estimates for tar sands reserves in Canada are 180 billion barrels and for Venezuela, 513 billion barrels.

The potential of tar sands as a replacement for conventional petroleum is limited to some extent by the expensive procedures required to prepare the material for a petroleum refinery. The material must first be converted from its solid or semi-solid state to a liquid. This step is accomplished in one of two ways: by strip mining the material and then heating it to a point at which it begins to flow, or by piping steam into tar sands bed and converting the bitumen in the beds to a liquid, a process known as *in situ* (Latin for "in place") recovery. The liquid obtained by either of these processes must then be purified, ridding it of contaminants that interfere with the refining process, such as the sand and dirt itself, water, metals, sulfur and nitrogen compounds, and solid particles of carbon. At this point in history, the cost of these processes exceeds the commercial value of most petroleum extracted by traditional methods. As the price of conventional petroleum increases, however, the demand for tar sands products has and will continue to increase.

As of early 2012, Canada is the only nation with a successful tar sands industry. That industry produces about one million barrels of oil per day from tar sands, which accounts for about 40 percent of that nation's output of petroleum. The largest amount of that oil is exported to the United States, which has only minimal tar sands reserves of its own (an estimated 12 to 19 billion barrels of oil, primarily in eastern Utah). Venezuela has not yet developed a robust tar sands industry because, according to some observers, of problems of development within the government-owned petroleum agency. No other nations are known at this point to have significantly large reserves of tar sands ("Oil Sands" 2012).

As the price of petroleum produced by conventional means increases, companies search for alternative methods of producing the fuel from other sources. Shale oil and tar sands are examples of these efforts. Two other technologies for increasing the supply of petroleum-like materials are compressed natural gas (CNG) and coal liquefaction. In one case (CNG), one fossil fuel, natural gas, is used as a substitute for another fossil fuel, petroleum. In the other case (coal liquefaction), another fossil fuel, coal, is converted into an entirely new form that mimics the physical and chemical properties of petroleum. Both of these technologies have current and potential value because they convert either natural gas or coal into the one fossil fuel that, with its apparently endless number of applications, by far dominates the world's energy equation.

CNG, as its name suggests, is produced simply by compressing natural gas at high pressures, typically about 210 atmospheres (3,100 pounds per square inch). The gas is then stored in spherical or cylindrical steel containers at these high pressures. Note that CNG is different from LNG, which is a *liquid* used for the transport of natural gas. By contrast, CNG is still a *gas,* although a gas at very high pressures. CNG has a number of advantages over gasoline in use with motor vehicles, primarily in that it burns more cleanly, producing a more efficient power production process and resulting in less pollution. CNG engines also tend to require lower maintenance, to result in less loss of fuel through leakage and evaporation, to extend the life of lubricating oils, and is less likely to auto-ignite. CNG also has its disadvantages, however, primarily because of the large size of containers needed to store the fuel in comparison with traditional gasoline tanks in a car. Also, in some instances, the fuel can be significantly more expensive than traditional gasoline, although that drawback tends to disappear as the cost of crude oil increases.

The first CNG vehicle (also known as a natural gas vehicle, or NGV) probably dates to the 1930s, during a period in which inventors were exploring a variety of alternative fuels for motor vehicles. CNG vehicles did not become popular at the time, primarily because the cost of gasoline was so low, but experimentation with the technology never completely ended. By the end of the 20th century, crude oil prices had risen sufficiently that natural gas vehicles once more seemed like a commercial possibility, and many major automobile producers began manufacturing cars, trucks, and buses that run on CNG. As of 2011, more than 12,670,000 natural gas vehicles were in operation around the world, an increase of 11.6 percent over the preceding year. More than 18,000 natural gas fueling stations were available for those vehicles, an increase of 10.2 percent over the 2009 numbers. The countries in the world with the largest number of natural gas vehicles were Pakistan (2,740,000 CNG vehicles), Iran (1,954,925), Argentina (1,901,116), Brazil (1,664,847), India (1,080,000), China (450,000), Colombia (340,000), and Thailand (218,459). As of March 2011, there were 112,000 natural gas vehicles and more than 1,000 fueling stations in the United States. Fuel for these vehicles cost $2.06 per gallon, compared to $3.69 for gallon for conventional regular gasoline ("Summary Data 2010" 2012).

Procedures for the conversion of coal to petroleum-like materials have been studied for more than a century. One of the earliest of these technologies was developed in the 1910s by German chemist Fredrich Karl Rudolf Bergius, who won the 1931 Nobel Prize in Chemistry for his study of high-pressure chemical reactions. The Bergius process is typical, in general, of other methods devised to convert coal to petroleum-like materials. Bituminous coal is first pulverized and then dried in a steam of hot gas. The pulverized coal is then mixed with heavy oil, which is a product of the reaction, and a catalyst is added to the mixture. The mixture is then heated in a reactor at temperatures of 750–930°F (400–500°C) and pressures of about

700 atmospheres (10,000 pounds per square inch). The reaction results in the formation of a variety of products, including heavy oils (which are then used in the Bergius reaction itself), middle oils, light oils, gasoline, and gaseous hydrocarbons. This mixture can be distilled in much the same way as is petroleum to produce fractions of various types.

As with other alternative types of fuels, the Bergius process and other methods for the conversion of coal to petroleum-like materials were unable to compete with petroleum from traditional sources for most of the 20th century. With the increase in crude oil prices toward the end of the century, however, that situation changed, and liquefaction of coal for the first time became a realistic economic alternative to traditional petroleum. Currently, there are a limited number of companies producing petroleum from coal in commercial quantities. The oldest of these facilities is Suid Afrikaanse Steenkool en Olie (South African Coal and Oil, or SASOL), founded in 1950. SASOL's coal-to-liquid (CTL) plant at Secunda currently produces about 4.5 million barrels of petroleum annually, about a quarter of the total South African demand for oil. That petroleum is used for the synthesis of other fuels and the generation of electricity. The company is also involved in other CTL projects in Canada, Germany, Iran, Italy, Mozambique, Qatar, and the United States, with additional plans for facilities in a number of other nations ("Welcome to Sasol" 2012).

The world's second commercial CTL plant opened for operation in Shenhua, in Majata Province, Inner Mongolia, in November 2010. The plant was built largely with technology developed in the United States, and has an operating capacity of 1.2 million tons (700,000 barrels) annually. The plant, and others like it planned for future construction in China, are designed to take advantage of China's immense coal reserves, more than a trillion tonnes, the largest in the world.

Little progress has been made toward the development of CTL plants in the United States. As of 2011, about a dozen projects are under consideration, although construction has not yet been started on any of them. The U.S. government National Energy Technology Laboratory estimates that construction may begin on the first three to five facilities by 2014, with a total of about 10 facilities under construction by the year 2017. In recent years the U.S. Congress has considered a number of bills that would provide financial support for such projects, although none of those bills have as yet passed either house (Cicero 2012).

References

Aliyev, Natig. "The History of Oil in Azerbaijan." http://azer.com/aiweb/categories/magazine/22_folder/22_articles/22_historyofoil.html. Accessed February 14, 2012.

Averitt, Paul. *Coal Resources of the United States, January 1, 1967.* Washington, D.C.: U.S. Government Printing Office, 1969. http://books.google.com/books?id=DyjwAAAAMAAJ&pg=PA60&lpg=PA60&dq=coal+production+in+the+united+states+1932+359+million+tons&source=bl&ots=Zjlf0dGVwR&sig=_VTevOH2dYROX4s2tmtOl- 0IZro&hl=en&sa=X&ei=PvI7T4H6KqOpiAKsm9CUDA&ved=0CFAQ6AEwCA#v=onepage&q=coal%20production%20in%20the%20united%20states%201932%20359%20million%20to-ns&f=false. Accessed February 15, 2012.

BP Statistical Review of World Energy June 2011. http://www.bp.com/assets/bp_internet/globalbp/globalbp_uk_english/reports_and_publications/statistical_energy_review_2011/STAGING/local_assets/pdf/statistical_review_of_world_energy_full_report_2011.pdf. Accessed February 15, 2012.

Cicero, Daniel C. "Coal-to-Liquids in the United States: Status and Roadmap." http://www.netl.doe.gov/technologies/hydrogen_clean_fuels/refshelf/presentations/CTL%20Tec%20Cicero%20June%2008.pdf. Accessed February 15, 2012.

"Coal." In *The 1911 Classic Encyclopedia*. http://www.1911 encyclopedia.org/Coal. Accessed February 14, 2012.

"Coal Mines in the Industrial Revolution." History Learning website. http://www.historylearningsite.co.uk/coal_mines_industrial_revolution.htm. Accessed February 14, 2012.

"Coal Production." BP. http://www.bp.com/assets/bp_internet/globalbp/globalbp_uk_english/reports_and_publications/statistical_energy_review_2011/STAGING/local_assets/spreadsheets/statistical_review_of_world_energy_full_report_2011.xls#' Coal—Production Mtoe'!A1. Accessed February 15, 2012.

Derry, T. K., and Trevor I. Williams. *A Short History of Technology from the Earliest Times to A.D. 1900*. New York: Dover Publications, 1993. (Reprint of the 1960 edition.)

Dyni, John R. *Geology and Resources of Some World Oil-shale Deposits*. Reston, VA: U.S. Geological Survey, 2006.

Gonzalez, Jorge. "Mexican Oil Industry." http://www.trinity.edu/jgonzal1/341f96g1.html. Accessed February 15, 2012.

Gray, Ralph D. *Indiana History: A Book of Readings*. Bloomington, IN: Indiana University Press, 1995.

"History." NaturalGas.org. http://www.naturalgas.org/overview/history.asp. Accessed February 14, 2012.

"History of Mine Safety and Health Legislation." Mine Safety and Health Administration. http://www.msha.gov/mshainfo/mshainf2.htm. Accessed February 15, 2012.

"History of Motor Car/Automobile Production 1900–2003." Carhistory4u. http://www.carhistory4u.com/the-last-100-years/car-production. Accessed February 15, 2012.

"History of Venezuela's Oil and Rentier Economy." Suburban Emergency Management Project. http://www.semp.us/publications/biot_reader.php?BiotID=476. Accessed February 15, 2012.

"King Coal." http://www.energyandhome.co.uk/page28.htm. Accessed February 15, 2012.

Kubiszewski, Ida. "Energy Timeline." http://www.earthportal.org/?p=80. Accessed February 14, 2012.

"Natural Gas: Proved Reserves." BP. http://www.bp.com/liveassets/bp_internet/globalbp/globalbp_uk_english/reports_and_publications/statistical_energy_review_2011/STAGING/local_assets/spreadsheets/statistical_review_of_world_energy_full_report_2011.xls#'Gas—Proved reserves'!A1. Accessed February 15, 2012.

"Natural Gas: The Golden Age of Gas?" Txchnologist. http://www.txchnologist.com/2011/from-basin-to-market-longest-natural-gas-pipelines. Accessed February 15, 2012.

Ngô, Christian, and J. B. Natowitz. *Our Energy Future: Resources, Alternatives, and the Environment.* Hoboken, NJ: Wiley, 2009.

"Oil Sands." Energy Minerals Division, American Association of Petroleum Geologists. http://emd.aapg.org/technical_areas/oil_sands.cfm. Accessed February 15, 2012.

Oosthoek, K.J.W. "The Role of Wood in World History." Environmental History Resources. http://www.eh-resources.org/wood.html#_edn13. Accessed February 14, 2012.

"Overview & FAQ." Baker Hughes Investor Relations. http://investor.shareholder.com/bhi/rig_counts/rc_index.cfm. Accessed February 15, 2012.

"Random Walks in the Low Countries." http://randomwalksinlowcountries.blogspot.com/2010/11/pneumatic-chemistry-in-markt-square.html. Accessed February 14, 2012.

Robinson, Paul R. "Petroleum Processing Overview." http://hogan.chem.lsu.edu/CHEM_1002/Notes/PetroleumProcessing.pdf. Accessed February 14, 2012.

"Shale Oil." Sultani Oil. http://www.sultanioil.com/shale-oil.php. Accessed May 30, 2012.

"Summary Data 2010." NGV Global. http://www.iangv.org/tools-resources/statistics.html. Accessed February 15, 2012.

"Sustainability in a Changing World." http://www.fao.org/docrep/T0829E/T0829E03.htm. Accessed February 14, 2012.

Thomson, Janet. *The Scot Who Lit the World: The Story of William Murdoch, Inventor of Gas Lighting.* Glasgow: Janet Thomson, 2003.

Vassilious, M. S. *Historical Dictionary of the Petroleum Industry.* Lanham, MD: Scarecrow Press, 2009.

"Welcome to Sasol." http://www.sasol.com/sasol_internet/frontend/navigation.jsp?navid=1&rootid=1. Accessed February 15, 2012.

"Whatever Happened to Standard Oil?" http://www.us-highways.com/sohist.htm. Accessed February 15, 2012.

Yergin, Daniel. *The Prize: The Epic Quest for Oil, Money, and Power.* New York: Simon and Schuster, 1991.

2 Problems, Controversies, and Solutions

The development of fossil fuel technology over the past two centuries has resulted in a profound revolution in the structure of human civilization. The use of coal, oil, and natural gas has made it possible for humans to heat and light their homes, office buildings, factories, and other structures efficiently, reliably, and at modest costs. Fossil fuels have also resulted in the development of many new forms of transportation that were unknown and unavailable to our ancestors: automobiles, trucks, buses, railroad engines, airplanes, and sailing ships of many kinds. Fossil fuels have also become the basis of an endless variety of industrial operations, from the manufacture of iron, steel, and other metals to the production of countless household products, to the generation of electricity for a vast array of purposes. In an entirely different field, fossil fuels have also become the raw materials from which a huge variety of synthetic materials—medicines, dyes, plastics, artificial fibers, packaging materials, paints, and so on—are now made. Indeed, in a way of which many people may be unaware, fossil fuels have revolutionized the materials of modern civilization nearly to the extent that it has changed the face of the world's energy equation.

No development of the scope of the fossil fuel age occurs, however, without concomitant problems and issues developing.

Natural gas rig in the Gulf of Mexico off the coast of Alabama. (iStockPhoto.com)

In the case of fossil fuels, two such issues dominate the field: the long-term availability of fossil fuels and the consequence of decreasing availability of such fuels, on the one hand, and the deleterious environmental effects resulting from the world's dependence on fossil fuels, on the other.

Peak Energy

The term *peak energy* refers to some point in history at which the production of a resource, such as coal, oil, petroleum, gold, zinc, or some other material, reaches a maximum. The presumption underlying the concept of a peak level of production is that most resources contained in the Earth's lithosphere, hydrosphere, and atmosphere are limited. They were created at some time in the Earth's history and are no longer being created. The quantities of these materials present on the Earth are, therefore, available for human use for only a limited period of time, until they are all used up, and then they will be gone forever. This scenario is of relatively limited value when talking about some natural resources, such as metals, because, at least in theory, metals can be recovered after first use, recycled, and re-used again . . . and again and again. Although there may be some ultimate limit to the number of times they can be re-used before they are essentially scattered throughout the environment, such a scenario is difficult to imagine.

Such is not the case with fossil fuels. Most experts believe that coal, oil, and petroleum were produced at some point in the distant past, probably at least 300 million years ago, under environmental conditions that no longer exist, did not exist in the more recent past, and are unlikely to exist in the foreseeable future. As a consequence, the amount of coal, oil, and natural gas now present on the Earth is probably all of those resources that humans will have at their disposal for as long as we can see into the future. As humans extract and burn up those resources, the total reserves that remain in the Earth continue to decrease until, presumably, none remain. At that point, humans will have to have found alternative forms of energy to replace coal, oil, and natural gas.

The trend in fossil fuel production and consumption can perhaps be broadly sketched from a common sense perspective: When a fossil fuel reserve is first discovered, equipment is produced for its extraction and production rates increase rapidly. As production increases, so does the demand for the product and its consumption. At some point, the product is being consumed more rapidly than it can be produced. Reserves of the product decrease as more is extracted to meet demand. As reserves decline, so does production. Eventually, reserves are exhausted, production slows dramatically and eventually comes to a halt, as must the consumption. It is easy to represent this scenario graphically by plotting annual production of the product on the *y*-axis of a graph against time (in years) on the *x*-axis of the graph. The shape of the graph would be familiar to most people: it is an almost-parabola with a rapidly ascending curve at the left side of the graph and a rapidly descending curve at the right side of the graph, with a maximum peak in between the two minimum points on the left and the right. This type of graph occurs frequently throughout mathematics and the sciences and is known as a normal curve, a Gaussian curve, a Cauchy curve, or a sine curve, although all of these curves are modestly different in shape from each other, primarily at the two ends of the curve.

The first person to use this form of analysis for fossil fuel supplies was the American geologist M. King Hubbert (1903–1989). In 1956 Hubbert presented a paper at a meeting of the American Petroleum Institute in San Antonio, Texas, in which he predicted that oil production in the lower 48 United States would peak between the late 1960s and early 1970s. He based this prediction on an analysis of the existing proven reserves in the United States, an estimate of possible future discoveries, and the current rate of oil production in the United States. Hubbert did not use any form of sophisticated mathematical analysis to obtain these results, although many geologists and mathematicians since his time have attempted to do so. In fact, a number of mathematical equations have been developed that closely fit the graph that

Hubbert included in his 1956 paper. The most common of those equations has the form:

$$Q(t) = \frac{Q_{max}}{1 + ae^{-bt}}$$

where Q_{max} is the total amount of oil available, $Q(t)$ is the cumulative production at some time, t, and a and b are constants (Laherrère 2012). (Don't worry about the math here; just recognize that experts have devised sophisticated methods for calculating the time at which peak supplies should occur for a resource.)

Hubbert argued that this pattern of production and peak oil should apply to any geographical entity, a single oil well, single oil field, a national oil supply, or worldwide oil supplies. He was convinced that the fundamental nature of fossil fuels, their once-in-history production, ensured that such supplies would eventually fail. Hubbert's paper was largely ignored or rejected out of hand at the time it was presented. At the time, most geologists were convinced that oil reserves were so immense and, to a large extent, still undiscovered, that talking about an end of fossil fuel reserves seemed absurd. Yet, in retrospect, Hubbert's predictions appear to be correct. For example, in retrospect, peak oil has already occurred in some parts of the United States, where maximum production has already occurred in states such as Pennsylvania (peak oil in 1891), Oklahoma (1927), and Texas (1972). In each of those states, as predicted, oil production has decreased since its peak oil year and never reached those maximum levels again ("World Annual Oil Production" 2012).

Similar patterns have been observed in many nations of the world, including the United States, where peak oil occurred in 1970, Venezuela (1970), Libya (1970), and Kuwait (1972). Table 2.1 lists countries for which peak oil years have been declared.

A number of countries appear to be some years away from peak oil. They include Algeria, Canada, Equatorial Guinea,

Table 2.1 Peak Oil Years for Various Countries

Country	Peak Production	2008 Production	Percentage Below Peak	Peak Year
United States	11,197	7,337	–35	1970
Venezuela	3,754	2,566	–32	1970
Libya	3,357	1,846	–45	1970
Kuwait	3,339	2,784	–17	1972
Iran	6,060	4,325	–29	1974
Indonesia	1,685	1,004	–41	1977
Romania	313	99	–68	1977
Trinidad & Tobago	230	149	–35	1978
Iraq	3,489	2,423	–31	1979
Brunei	261	175	–33	1979
Tunisia	118	89	–25	1980
Peru	196	120	–39	1982
Cameroon	181	84	–54	1985
Russian Federation*	11,484	9,886	–14	1987
Egypt	941	722	–23	1993
Syria	596	398	–33	1995
Gabon	365	235	–36	1996
Argentina	890	682	–23	1998
Colombia	838	618	–26	1999
United Kingdom	2,909	1,544	–47	1999
Uzbekistan	191	111	–42	1999
Australia	809	556	–31	2000
Norway	3,418	2,455	–28	2001
Oman	961	728	–24	2001
Yemen	457	305	–33	2002
Mexico	3,824	3,157	–17	2004
Malaysia	793	754	–5	2004
Vietnam	427	317	–26	2004
Denmark	390	287	–26	2004
Nigeria	2,580	2,170	–16	2005
Chad	173	127	–27	2005
Italy	127	108	–15	2005

* Comparisons are difficult because Russian Federation was preceded by the Soviet Union, a different collection of national states.

Source: Adapted from "Is Peak Oil Real?" (2012). Produced from data in *BP Statistical Review of World Energy 2009.* Used by permission.

India, Malaysia, and Saudi Arabia, for all of whom oil production has decreased by modest amounts (1 to 5%) over their peak years, but for whom it is too early to say that peak oil has passed. In fact, peak oil can be identified only when that point has actually been reached a few years in the past and a real decline in production can be seen with little or no uncertainty. Countries for whom peak oil appears to be almost certainly some years into the future include Angola, Azerbaijan, Brazil, China, Kazakhstan, Qatar, Sudan, Thailand, Turkmenistan, and United Arab Emirates. Overall, as of the end of 2009, about 60 percent of the world's oil production appears to have come from nations whose peak oil year has passed, or where oil production is currently flat and not increasing. The remaining 40 percent comes from the nations that are still experiencing growth in the production of petroleum. World oil production appears already to have peaked. In its annual *World Energy Outlook* for 2010, the International Energy Agency (IEA) concluded that "[c]rude oil output reaches an undulating plateau of 68–69 mb/d [million barrels per day] by 2020, but never regains its all-time peak of 70 mb/d reached in 2006" (International Energy Agency 2012, 6).

It should be noted that predictions or announcements of peak oil are always subject to unexpected events, such as the Arab oil embargo of 1967, 1973, and 1979, when political events interfered with the normal operation of the market and drastically reduced the production of oil. Most authorities believe that such temporary disruptions, however, are only blips in the long-range trend for petroleum production and have no long-term effects on the existence or general timing of peak oil.

Other Views about Peak Oil

The theory of peak oil has a number of prominent critics. The most common argument proposed by these critics is that peak oil proponents vastly underestimate the amount of oil still present beneath the Earth's surface and that improved methods of finding and removing that oil will extend supplies far beyond any estimates currently being made. They point out that oil exploration will result in the discovery of new oil fields and that the reserves in existing fields are far greater than have so far been estimated. In addition, they point to some important new technologies, such as tar sands, shale oil, and biomass, which represent huge new reserves of petroleum not so far counted by peak oil proponents. Finally, they suggest that market forces will drive petroleum companies to find, extract, and make available larger supplies of oil than they have thus far produced.

One of the most articulate spokespersons for this position is Daniel Yergin, chairman of IHS Cambridge Energy Research Associates, an energy research and consulting firm. Yergin makes many of these arguments and points out that experts have predicted the arrival of peak oil at least four times in the past: first in the 1880s, when all U.S. oil was being produced in Pennsylvania and those resources began to run low; again after both World War I and World War II when reserves in Oklahoma and Texas began to be exhausted; and a fourth time in the 1970s. The problem with peak oil theories, according to Yergin, is that every time concern develops about running out of oil, companies use new technologies to find and extract oil, thus continuing to increase the supply of the fuel. In a recent essay, he said that "[i]n the oil and gas industry, technologies are constantly being developed to find new resources and to produce more—and more efficiently—from existing fields," so that we can always count on there being new resources of oil for the future. He concludes his essay by observing that, "Things don't stand still in the energy industry. With the passage of time, unconventional sources of oil, in all their variety, become a familiar part of the world's petroleum supply. They help to explain why the plateau continues to recede into the horizon—and why, on a global view, [M. King] Hubbert's Peak is still not in sight" (Yergin 2012).

Representatives from the petroleum industry are among the strongest critics of the peak oil theory. At a meeting on energy supplies held in Davos, Switzerland, in January 2010, for example, Khalid A. Al-Falih, president and CEO of Saudi Aramco, the world's largest oil company, said that there is still plenty of oil in the ground, and the world should not worry about running short of the fuel. "Of the 4 trillion (barrels) of oil the planet is endowed with," he said, "only 1 has been produced." "Granted most of what remains is more difficult and complex (to exploit)," he continued, "there's no doubt we can do a lot more than the 95,100 (million barrels) that are projected in the next few decades" ("Davos" 2012).

In the end, the debate does not appear to be quite one as to *whether* there is a time when peak oil occurs, but *when* it might appear. Critics appear to be saying that so much oil remains in the Earth that humans do not have to fear about running out of the resource at any foreseeable time in the future. Proponents of peak oil disagree and are generally quite willing to set a date at which that scenario will occur, often in the very distant future.

Economists also point out the importance of market forces in determining when (although not necessarily whether) peak oil will occur. They say that a combination of demand for oil and a political will to satisfy that demand can make changes in the availability of oil in the marketplace. For example, the United States has large petroleum reserves that are so far untouched, or nearly so, in places such as the Arctic National Wildlife Refuge and a number of offshore sites. It also has the potential for producing oil from massive amounts of oil shale in the Green River Basin of Wyoming and from coal-to-liquids plants in parts of the West. But mining of these sites has been delayed by a variety of factors, such as concerns about environmental damage. If they could be put into operation, however, oil production in the United States might begin to soar well beyond the maximum level reached in the early 1970s, when the nation supposedly reached its peak oil moment.

The usual response to this argument by proponents of peak oil is that market forces do not always work in the same way with resources that are in limited supply, such as fossil fuels, as they do with other products. The supply of oil cannot be increased by market forces simply by increasing demand, as is often the case, if there is not enough oil in the ground to meet that demand.

No resolution between the pro and con peak-oil forces appears to be possible. It may simply be a matter about which time itself will tell, whether, a decade or two from now or farther into the future, oil supplies continue to dwindle and the world is indeed in the downside of a peak oil scenario or whether oil companies continue to respond to ever-increasing demands for this essential fuel.

Peak Coal and Peak Natural Gas

An argument can be made for the existence of a period of peak coal and peak natural gas similar to that for peak oil, and, in fact, it has often been made. Again, the point is that there are only limited supplies of these materials and that, at some point, reserves and production will no longer be able to keep up with demands. Since the arguments for and against this proposition are essentially the same as they are for peak oil, they will not be repeated here. A review of the available evidence may be useful, however.

As with peak oil, estimates as to when peak coal or peak natural gas will occur vary widely across the spectrum, often depending to a considerable degree on the person or agency making the estimate. In general, energy companies appear to take a more optimistic view of the situation, placing peak coal and peak natural gas, even if they admit that such scenarios are possible, many years or decades into the future. For example, the World Coal Association (WCA) argues that peak coal is not an issue about which the world needs to be concerned. On its website, the association notes that the reserves-to-production (*R/P*) ratio has been falling, which one would normally construe to mean movement toward peak coal. Such is not the case, however, WCA says, because companies have not felt any incentive to extend their exploration for new reserves or extraction from existing sites. The association concludes that "[t]here is no economic need for companies to prove long-term reserves," and that "[c]oal reserves could be extended further through a number of developments including . . . the discovery of new reserves through ongoing and improved exploration activities [and] advances in mining techniques, which will allow previously inaccessible reserves to be reached" (World Coal Association 2012).

A number of governmental, intergovernmental, industrial, academic, and special interest groups have attempted to estimate peak coal scenarios for individual coal mining areas, specific nations, and the world overall. Those estimates vary from a few years to more than a hundred years into the future for the arrival of peak coal. Of course, some regions have already reached peak coal. For example, the United Kingdom reached peak coal sometime around World War I, with maximum production of nearly 330 million tons per year. That number dropped throughout the rest of the 20th century and reached less than 10 million tons in the early 2000s. Similarly, the German coal industry reached its peak production the years just before and just after World War II at about 165 million tons annually. It then fell off to less than 10 million tons annually also in the early 2000s. In the United States, coal production in the state of Pennsylvania, once the largest-producing state in the nation, reached peak coal in about 1910 at more than 250 million short tons per year, after which it declined to about 60 million short tons annually at the end of the 20th century (Energy Watch Group 2012, 42; Rembrandt 2012b, Figure 1).

In an important review of world coal resources in 2007, the Energy Watch Group attempted to estimate the point at which peak coal would occur in a number of nations. They predicted that that event would occur in China in about 2020,

India in about 2025, South Africa and Australia in about 2035, the United States in about 2040, and Russia in about 2070. Overall, they estimated that the world would reach a peak coal scenario sometime between 2020 and 2050. The difference in these estimates reflects the most optimistic set of circumstances contributing to the estimate (the 2050 estimate) and the standard set of circumstances (life as usual; the 2020 estimate) (Energy Watch Group 2012).

In comparison with the Uppsala-Systemtechnik estimates, the Energy Watch Group (EWG) predicted that global coal production would peak around 2025 in the best case scenario. EWG also estimated that peak coal would be reached in the United States more than 200 years in the future, in China in about 2020, and in Canada, sometime between 2030 and 2040 (Energy Watch Group 2012). (This report focused on the situation in four countries: the United States, Canada, China, and Germany.) As with oil and coal, experts differ widely on the possibility and/or timing of a peak natural gas scenario. Many oil and gas executives acknowledge that new discoveries of natural gas reserves have already diminished considerably, with peak gas discoveries dating perhaps to about 1960. Since that time, the demand for natural gas has continued to grow significantly, whereas new gas discoveries have plunged to some of the lowest levels in history. Some data suggest that peak gas scenarios may already have occurred in some countries. Gas production in Italy, for example, peaked at about 670 billion cubic feet annually in 1998, and has since fallen somewhat dramatically to less than 350 billion cubic feet per year. A similar trend has occurred in the United Kingdom, where gas production peaked at just less than 4.0 trillion cubic feet per year in 2000 and has since dropped to about 2.0 trillion cubic feet per year in 2010. The United Kingdom became a net importer of natural gas for the first time in 2004. In the United States, the production of natural gas has fluctuated between about 18 and 21 trillion cubic feet for four decades, with current levels at the high end of the range (20,159 billion cubic feet in 2008;

20,580 billion cubic feet in 2009; and 21,577 billion cubic feet in 2010) ("Italy Needs to Face up to the Future" 2012; Rembrandt 2012a; "Annual U.S. Natural Gas Marketed Production" 2012).

Not surprisingly, some observers are not convinced that the world has reached, or even approached, a peak gas scenario. As usual, the argument for this position tends to be that (1) new natural gas fields are yet to be discovered and exploited and (2) only a small fraction of available reserves have thus far been extracted from existing fields. For example, blogger and oil and gas worker Nolan Hart has taken the position that new technology in the oil and gas industry will vastly increase the supplies of both fuels and that peak gas simply will not occur. In a March 2010 blog, he wrote that, "[w]e suddenly have over a one hundred year supply of natural gas at current consumption rates and that number has been growing by about one decade more each year since 2005." He goes on to claim that, "scientists estimate there is a nine thousand year supply, at current consumption rates, lying frozen on the seafloor off of the Atlantic coast of the United States. All we need to do is build some high tech underwater Roombas to go down there and vacuum it up. No, peak natural gas is not coming anytime soon" (Hart 2012). (No citations are provided for these data, so they should be taken at face value.) Two other estimates for peak gas for the world are available from the U.S. Energy Information Administration (EIA), which predicts a continued increase in gas production worldwide until at least 2030, and from the Joint Transport Research Centre at Uppsala University, which estimates a peak gas scenario sometime around 2015–2020 (U.S. Energy Information Administration 2012b; Aleklett 2007).

Extending Fossil Fuel Supplies

For those who are convinced of the existence of peak energy, the challenge is to find ways of conserving our present energy supplies, so as to possibly delay to some extent the arrival of peak oil, peak coal, and peak gas, and to find alternative energy sources that can replace our falling supplies of fossil fuels. For such individuals, it is essential to explore the options presented by biofuels, solar energy, wind power, geothermal resources, tidal and wave energy, nuclear power, and other alternative and often renewable sources of energy. This challenge is discussed later in this chapter.

For those who are dubious about the concept of peak energy or who, at least, think peak oil, peak coal, and peak gas are located at some distant time in the future, the challenge is different; it is to find ways of extending existing reserves of coal, oil, and gas to ensure that production can continue to grow to meet growing demands for these fuels. Here are some of the specific ways of dealing with that challenge.

Exploitation of Known Reserves

One of the most obvious ways of increasing petroleum production is to begin extraction of fossil fuel resources from reserves that have already been identified, but not yet exploited. An example of such a possibility can be found in the Arctic National Wildlife Refuge (ANWR) in the northeastern corner of the state of Alaska. ANWR was created in the Alaska National Interest Lands Conservation Act of 1980, which set aside more than 100,000,000 acres (420,000 square kilometers) of national parks, wildlife refuges, and wilderness areas from federal holdings in the state. Drilling for oil and gas was specifically permitted in a region of ANWR known as Section 1002 along the northernmost coast of the refuge, adjacent to Prudhoe Bay, but not without expressed prior approval by Congress for such exploration. Over the next three decades, debates between proponents of oil exploration in the area and environmentalists have continued with the primary issue being the value of possible oil reserves in Section 1002 versus potential environmental damage, especially to an area that is the primary calving area for Alaskan caribou. Proponents of drilling have argued that exploitation of petroleum reserves in Section 1002 could significantly improve the United States' energy equation, with a daily production of oil that would equal all the oil imported from Saudi Arabia. In 2011, for example, Representative Doc Hastings (R-WA) submitted legislation that would open Section 1002 to exploratory drilling. He argued that drilling would "create thousands of jobs, generate billions in new revenue and help reduce our dependence on foreign sources of oil." Exploitation of Section 1002 would presumably also delay the problems posed by peak oil in the United States. Rep. Hastings based his argument on a 2005 U.S. Geological Survey report estimating the total oil reserves in Section 1002 at 10.4 billion barrels of oil, "more than the known oil reserves of entire countries that the U.S. currently imports oil from, including: Mexico, Angola, Azerbaijan, Norway, India, Indonesia, Malaysia, Egypt, Australia and New Zealand, Turkmenistan, and Uzbekistan," according to Rep. Hastings (Alaska Energy for American Jobs Act 2012).

The U.S. EIA has studied the potential effects of drilling in Section 1002 on the nation's energy equation. Its 2008 report assumed three possible scenarios, one based on the highest reserves imaginable (the high estimate), the lowest reserves imaginable (the low estimate), and an average of these two estimates (the mean estimate). EIA analysts then calculated that crude oil production would actually begin in 2018 leading to a maximum output of 780,000 barrels per day in 2027 under the mean estimate, 510,000 barrels per day in 2028 under the low estimate, and 1,450,000 barrels per day in 2027 under the high estimate. Production would then begin to fall off, as in any peak oil scenario. Additional petroleum resources from ANWR would, EIA estimated, reduce oil imports by the United States from between 8 (low estimate) and 12 (high estimate) percent (U.S. Energy Information Administration 2012a).

Another resource that has not been fully exploited, according to those who doubt the theory of peak oil, is offshore drilling. Again they argue that huge reserves of petroleum and natural gas are available under the oceans, many of which have not been exploited because of environmental and other concerns. Table 2.2 shows the EIA estimate of the amount of oil and natural gas available from offshore sources on the continental shelf surrounding the United States.

The American Petroleum Institute (API) has provided one of the clearest and most succinct explanations as to why exploration for oil and natural gas should be allowed on both federal lands, like ANWR, and on offshore sites. In a policy statement on the issue, the API points out that "[o]il and natural gas from federal lands and waters is critical to meeting the nation's energy needs, providing approximately 30 percent of all oil and

Table 2.2 Technically Recoverable Resources of Crude Oil and Natural Gas in the Outer Continental Shelf, as of January 1, 2007

Resource Area and Category	Crude Oil*	Natural Gas†
Proved Reserves		
Gulf of Mexico	3.66	14.55
Pacific	0.44	0.81
Atlantic	0.00	0.00
Alaska	0.03	0.00
Total Proved Reserves	4.13	15.36
Inferred Reserves		
Gulf of Mexico	9.33	48.83
Pacific	0.89	0.26
Atlantic	0.00	0.00
Alaska	0.00	0.00
Total Inferred Reserves	10.21	19.09
Undiscovered Resources		
Gulf of Mexico	37.94	204.67
Pacific	10.50	18.43
Atlantic	3.92	36.50
Alaska	28.61	132.06
Total Undiscovered Resources	78.97	391.66
Total Reserves, All	**93.31**	**456.11**

*Billion barrels.
†Trillion cubic feet.

Source: "Impact of Limitations on Access to Oil and Natural Gas Resources in the Federal Outer Continental Shelf" (2011).

38 percent of all natural gas produced in the United States. . . . Greater access to these areas is needed because that's where the remaining oil and natural gas accumulations are likely to be located—particularly the larger ones" ("Why We Need More Development" 2012).

The U.S. federal government first became involved in the regulation of offshore oil and gas drilling in 1953 when Congress passed the U.S. Submerged Lands Act, which claimed mineral rights to all parts of the continental shelf beyond a distance of three miles from shore. (Individual states, cities, and other governmental bodies have authority over parts of the ocean bottom within those three miles.) In 1983, President Ronald Reagan further defined the parameters of offshore drilling by setting the outer limit over which the government intended to exert control to a distance of 200 miles from the shore. Later Congressional action specifically regulating offshore drilling operations was usually based on specific disasters, such as the Santa Barbara oil spill of 1969, or extreme fluctuations in oil and gas prices, such as those that occurred as the result of the Arab oil embargoes of 1967, 1973, and 1979.

In 1982, the Congress passed the first comprehensive limit on offshore drilling by setting aside 763,000 acres of the continental shelf off the coast of California. Over the next decade, that type of ban was extended to other regions in the Gulf of Mexico and along the Atlantic coast. Finally, in 1990, President George H. W. Bush issued an executive order establishing a moratorium on offshore drilling across these three general areas, a moratorium that was extended by President Bill Clinton to 2012.

The trend toward protecting the country's coasts and offshore waters was reversed in 2008 when President George W. Bush opened wide tracts of offshore regions to oil and gas drilling. Although incoming President Barack Obama had campaigned against further expansion of offshore drilling, he instead followed Bush's lead and, in March 2010, extended the regions in which offshore drilling was to be permitted. Only a month later, a fire at BP's *Deepwater Horizon* drilling rig in the Gulf of Mexico resulted in the worst oil spill in history, causing Obama to change his mind about his newly announced plans. He instead imposed a moratorium on all drilling in the Gulf until all offshore rigs could be inspected, a ban that was almost immediately overturned by the courts. Since that time, the federal government has oscillated back and forth with regard to its offshore drilling policies, attempting to strike a balance between efforts to increase oil and gas production from U.S. territories with concerns over possible environmental consequences. The most recent version of the federal government's policy on offshore drilling was announced in November 2011. As provided for in the Proposed Outer Continental Shelf Oil and Gas Leasing Program for 2012–2017, more than 75 percent of recoverable oil and gas resources in federal offshore areas will be available for exploration and development by private companies until 2017. A total of 15 lease sales will be available under the program, including 12 in the Gulf of Mexico and three off the coast of Alaska.

This review of U.S. policy with regard to offshore drilling should by no means suggest that the United States is the only nation in the world dealing with issues of offshore drilling. Indeed, as shown in Table 2.3, offshore drilling is carried out in many parts of the world and the debates over how, when, and where that technology is to be used differs greatly depending on local governmental policies, economics, technology, environmental concerns, and other issues. As one of many possible examples of the problems related to offshore drilling was the May 2010 announcement by Royal Dutch Shell that it would discontinue offshore drilling in the Niger Delta of Nigeria, a region not unlike the Mississippi River Delta in Louisiana. The company's operations had been plagued by a number of spills and other disasters—including more than 2,000 in 2009 alone—that had led to increasing scrutiny by the government of Nigeria of the company's activities, especially in light of the *Deepwater Horizon* disaster a month earlier. The company

Table 2.3 Offshore Drilling Rigs Worldwide

	Drilling Rigs*	
Region	**Available**	**Under Contract**
U.S. Gulf of Mexico	115	67
South America	135	109
Europe or Mediterranean Sea	117	104
West Africa	69	55
Middle East	120	99
Asia or Australia†	153	126
Worldwide	820	656

*Rigs under contract are those that are currently in active use in a drilling project, being moved from place to place, undergoing maintenance, or in some other ready-to-operate condition.
†Excluding India.

Source: Weekly Rig Count (2011).

announced that it would invest $8 billion in 2012, and $1 billion annually over the next decade, to restore damage done to the Niger Delta as a result of drilling operations. It also established a $4 billion fund for compensation for perceived injustices perpetrated by the company's operations ("The Comprehensive Shell Remediation Plan" 2012).

Tar Sands and Shale Oil

As mentioned in Chapter 1, tar sands and shale oil represent perhaps the greatest hope for peak energy naysayers' hope for a fossil-fuel-rich future. In a recent (November–December 2011) article about tar sands in Canada's Alberta Province, a headline asked "What Peak Oil?," suggesting that this resource was so vast that talk about running out of oil seemed almost ludicrous. And, on the surface, the numbers seem impressive. Canada's total bitumen reserves are currently estimated at 169 billion barrels, compared to conventional petroleum reserves in the country of about 1.5 billion barrels. Those figures place Canada (and, more specifically, Alberta), second in the world in proven oil deposits, behind only Saudi Arabia, and ahead of such oil giants as Iraq, Venezuela, Russia, and the United

States in proven reserves. They also encourage peak oil deniers to argue that concerns about dwindling reserves of petroleum around the world are vastly overstated (Fairley 2011). For example, one of the world's best-known writers on the topic of peak oil, Daniel Yergin, was still writing as late as September 2011 that supporters of peak oil theories had vastly underrated the contribution that alternative fossil fuel resources, such as tar sands, could make to solving the world's energy equation. "Meeting future demand will require innovation, investment and the development of more challenging resources," he wrote in the *Wall Street Journal.* But that innovation is certainly possible, he went on, now that oil previously regarded as "inaccessible or uneconomical," such as the "vast oil sands of Canada," are part of the energy equation (Yergin 2012).

The extraction and use of bitumen is hardly a new phenomenon in the 21st century. The material has been used for millennia for a variety of purposes, such as the calking of wood planks in buildings. Explorers in the Canadian West knew about and commented on the vast reserves of bitumen in the tar sands of the Athabasca region of Alberta as early as 1719. Natives in the areas had been using the material for even longer periods of time, for example, in waterproofing their canoes. No efforts were made to develop these resources commercially, however, until the late 1960s with the opening of the Great Canadian Oil Sands plant (now Suncor Energy) about 20 miles north of Fort McMurray in 1967. Since that time, the industry has grown substantially, as shown by production figures in Table 2.4.

Most experts believe that the recent growth in production from the Athabasca oil sands field is just the beginning of a boom in the industry. A November 2011 report by Canada's National Energy Board predicted that oil sands output in Alberta would triple by the year 2035 (*Canada's Energy Future* 2011, ix). One consequence of this anticipated increase has been the development of plans to construct a 1,700-mile-long pipeline from Alberta to Texas for delivery of crude oil for

Table 2.4 Production of Bitumen in Canada, 1967–2010

Year	Production*	Reserves[†]
1967	42	
1968	64	
1969	69	
1970	38	
1971	15	
1972	71	
1973	71	
1974	73	
1975	199	
1976	440	
1977	486	
1978	453	
1979	581	
1980	556	
1981	753	
1982	1,281	7,050
1983	1,455	6,320
1984	1,943	6,370
1985	3,030	136,507
1986	5,410	147,760
1987	6,731	151,415
1988	7,546	221,997
1989	7,474	218,995
1990	7,856	174,591
1991	7,113	164,324
1992	7,362	164,725
1993	7,685	158,844
1994	7,810	169,640
1995	8,621	197,540
1996	9,505	211,105
1997	13,806	229,837
1998	16,364	220,514
1999	14,171	248,058
2000	16,781	286,826
2001	17,954	289,227
2002	17,560	321,745
2003	20,242	322,999

(continued)

Table 2.4 *(continued)*

Year	Production*	Reserves†
2004	22,459	330,868
2005	25,553	393,253
2006	27,161	747,891
2007	29,230	786,590
2008	32,021	1,056,864
2009	33,047	1,057,794
2010	40,823	n/a
Total	**421,904**	

*Thousand cubic meters.
†Thousand cubic meters; at end of the year.

Source: Canadian Association of Petroleum Producers (2011, Tables 2.10.1a and 3.2a)

refining. The plan received a setback in November 2011 when the U.S. government decided to put approval for the pipeline on hold for 18 months of additional consideration. The U.S. decision was based on concerns about possible environmental consequences of the pipeline project in the eight states through which it would run. (For an update on this constantly changing issue, see, e.g., Winston [2012].)

The enthusiasm among peak oil deniers about the contribution of oil sands to the world's energy equation has been modified somewhat by a host of problems associated with mining for the product. One of those problems has been unexpectedly large increases in the extraction of crude oil from tar sands. By the mid-2000s, companies were reporting that the cost of extracting crude oil by mining technology had increased by as much as 32 percent in a single year (2006) and the cost of *in situ* removal had increased by 26 percent during the same period. The primary reasons for these increases were larger-than-expected labor costs and the cost of equipment.

Water use is another matter of concern associated with crude oil recovery from tar sands. About 12 barrels of water are needed to recover a single barrel of crude oil from tar sands. About 80 percent of that water can be recycled, but that still leaves anywhere from two to four barrels of water per barrel of crude oil left behind in the environment. Since that water is often contaminated with mining wastes, it poses a long-term problem of environmental cleanup.

One of the most problematic restrictions on bitumen production, ironically, is potential energy shortages. About 1,000 cubic feet of natural gas must be burnt in order to extract one barrel of bitumen from tar sands. The investment of that amount of energy in producing an alternative form of energy is justified as long as the per-barrel price of bitumen is high enough, which it currently is. However, the question is whether Canada has the natural gas reserves that will make possible production of bitumen at the levels that have been predicted by government and industry analysts. Some experts think not. According to some estimates, the production of the quantities of bitumen usually predicted for the next decade might actually require all of the natural gas available in Canada. As one authority, Kjell Aleklett, at Sweden's Uppsala University, has noted, "The supply of natural gas in North America is not adequate to support a future Canadian oil sands industry with today's dependence on natural gas" (Aleklett 2007).

A third issue associated with the development of oil sands reserves in Alberta is the potential environmental impacts of mining operations. Most of the tar sands exploitation occurs in regions of Alberta covered by boreal forest and muskeg land. (Muskeg is a bog-like land formation common in boreal regions.) Canadian law requires mining companies to restore land to its previous condition after mining has been completed in an area, and mine companies have explained that they are, to a large extent, meeting that requirement. In 2011, the Syncrude Corporation announced that it restored 4,500 hectares, or about 3.2 percent, of all the land mined thus far. Of that amount, 3,400 hectares had been completely reclaimed and returned to its natural state and an additional 1,000 hectares had been capped with soil. The company pointed out that it had received its first certificate of reclamation in 2008 covering a 104-hectare area known as Gateway Hill ("About Syncrude" 2012).

Independent observers were not so certain about the success of the reclamation conducted by Syncrude. The Parkland Institute, a research organization associated with the University of Alberta, noted that it had taken Syncrude 10 years to reclaim the 104-hectare (about one square kilometer) piece of land and that reclamation of the area meant primarily that the company had dumped overburden from other mining areas on the previously mined region, not really restoring it to its original condition.

Critics of tar sands mining are also concerned about the contribution of the technology to the release of greenhouse gases. (The term *greenhouse gases* refers to gases that absorb heat in the Earth's atmosphere, thereby contributing to increases in the Earth's average annual temperature. The most important of the greenhouse gases are carbon dioxide, methane, water vapor, nitrous oxide, and ozone.) The very large consumption of methane gas in producing bitumen has significantly increased Canada's contribution to the amount of carbon dioxide released to the atmosphere annually. Government officials are under the impression that this mining technology is a major reason that Canada will not be able to meet its commitments under the Kyoto Protocol to reduce greenhouse gas emissions by 6 percent below 1990 levels by 2012 and by 17 percent below 2005 levels by 2020. Perhaps reflecting its concerns about falling behind on the Kyoto commitment, the Canadian federal government has reduced its target for greenhouse gas emissions from 310 million tons in 2007 to 31 million tons in 2011.

The only other place in the world where tar sands exist in commercially exploitable quantities is the Orinoco Belt, which runs along the coast of Venezuela. The region covers an area of about 21,000 square miles (55,000 square kilometers) and contains an estimated 1.2 trillion barrels of petroleum-like product, an amount thought to be about equal to all other conventional petroleum reserves in the world, and about 20 percent smaller than Canadian tar sands reserves. The product obtained from the Orinoco Belt is somewhat different in physical properties from the bitumen obtained from Canadian resources. In particular, it is denser and less viscous than the Canadian product, explaining the tendency to refer to the Venezuelan product as *extra-heavy crude,* rather than tar sands, oil sands, or bitumen. In 2011, the U.S. Geological Survey estimated that about 513 billion barrels of Orinoco heavy crude are recoverable by current technology ("Heavy Oil Resources" 2012).

Exploration of Venezuela's extra-heavy crude reserves began in 1935 with the drilling of the first well in the area, La Canoa 1. The well produced 40 barrels of oil per day for a short period of time, but no further research in the area was done for four decades. Then, the federal government authorized a new study of the area by a joint team from Corporación Venezolana del Petróleo (CVP), the state petroleum corporation, and the Ministry of Mines and Hydrocarbons. The research team drilled 669 wells between 1979 and 1983 which, by the end of 1985, were producing 80,000 barrels of extra-crude oil per day. Exploration and development of the Orinoco resources essentially came to a halt in 1986 with the discovery of the oil fields around Maracaibo Lake that made the extraction of extra-heavy crude economically impractical. The reserves lay essentially untouched then until 2005, when President Hugo Chavez announced a plan to open the Orinoco Belt to development by consortia of companies approved by the government. For a host of reasons, that plan had still not been realized as of early 2012, with no exploration initiated nor, in most cases, any actual agreements signed for companies to initiate an Orinoco-based project (Toro 2012).

Reflecting these problems, Petróleos de Venezuela, S.A. (PDVSA) the Venezuelan state-owned petroleum company, has regularly modified and delayed its projections as to when oil from the Orinoco Belt will begin flowing and, once it starts to flow, how much will be produced annually. Its most recent

(2010) projection was for the production of 4.46 million barrels a day by 2015 from the Orinoco Belt, down from a projection of 4.9 million barrels a day by 2013 in its 2009 projection. One blogger on the subject has suggested that it might be a good idea to save time just by republishing the headline "PDVSA cuts and delays production goals, raises capex forecast," every year. Under these circumstances, the objections that have already been raised to mining for extra-heavy crude in the Orinoco Belt, very much like those raised with regard to mining for bitumen in Alberta, are somewhat moot ("PDVSA Cuts and Delays Production Goals" 2012).

As with tar sands, the growing availability of shale gas and shale oil is prompting some observers to suggest that the public have second thoughts about, and not be so concerned over the prospects of, peak oil and peak gas. As one recent headline in the United Kingdom's *The Guardian* newspaper announced, "The Peak Oil Brigade Is Leading Us into Bad Policymaking on Energy. One can't assume energy prices are going ever upwards. The real problem is there may be too much fossil fuel, not too little." The argument of the article was that there is plenty of fossil fuel still stored underground. The problem is to find technologically efficient ways to extract those resources at affordable prices. The writer seemed convinced that such an option was possible with alternative fossil fuels, such as shale gas and shale oil (Helm 2012).

Certainly, the gross reserve numbers for shale gas and shale oil should give one pause. According to the World Energy Council's *2010 Survey of Energy Resources,* the world's largest reserves of shale oil by far are found in the United States, with a total of 3,706,825 million barrels of proved reserves, 77 percent of the world total. The next largest reserves are located in China (354,430 million barrels) and the Russian Federation (247,883 million barrels), with no other nation having more than 100,000 million barrels of reserves (*2010 Survey of Energy Resources* 2012, 101). Comparable data for shale gas reserves are shown in Table 2.5.

Table 2.5 Technically Recoverable Shale Gas Resources by Country (Trillion Cubic Feet)

Country	Reserves
Algeria	231
Argentina	774
Australia	396
Bolivia	48
Brazil	226
Canada	388
Chile	64
China	1,275
Colombia	19
Denmark	23
France	180
Germany	8
India	63
Libya	290
Lithuania	4
Mexico	681
Morocco	11
Netherlands	17
Norway	83
Pakistan	51
Paraguay	62
Poland	187
South Africa	485
Sweden	41
Tunisia	18
Turkey	15
Ukraine	42
United Kingdom	20
United States	862
Uruguay	21
Venezuela	11
Western Sahara	7
Total	6,622

Source: World Shale Gas Resources (2011, Table 1, 4).

One of the most discussed reservoirs for shale oil and shale gas is the so-called Bakken Formation, which underlies an extensive range of land in Montana, North Dakota, South Dakota, Alberta, and Saskatchewan. The formation is thought to extend over an area of about 200,000 square miles (520,000 square kilometers). According to a 2008 estimate by the U.S. Geological Survey (USGS), the formation may contain anywhere from 3.0 to 4.3 billion barrels of recoverable oil, making it one of the richest petroleum reserves in the world. In addition, the formation is thought to contain as much as 1.85 trillion cubic feet of natural gas and 148 million barrels of natural gas liquids ("3 to 4.3 Billion Barrels" 2012). (The term *natural gas liquid* refers to low molecular weight hydrocarbons that can be liquid under natural conditions, such as propane and the isomers of butane, pentane, and hexane.) Oil in the Bakken Formation resides in three horizontal layers with a total thickness of up to 130 feet (40 meters). The middle layer is moderately porous, allowing the use of conventional vertical drilling techniques for extraction. The upper and lower layers consist of shale rock that is not very porous or permeable, such that oil typically does not pool in concentrations large enough to permit conventional vertical drilling technology. (Strata in which grain size is small, or the space between grains is small, making it difficult for oil and gas to penetrate into the strata, are said to be tight fields.) Overall, the site presents serious challenges to efforts for its removal. These challenges are reflected in the fact that production from wells in the formation never exceeded much more than 10,000 barrels per day between 1973 and 1991, and often dropped to much less than that during the period. Production began to increase significantly in January 2003 with the introduction of new technology, especially horizontal drilling and hydraulic fracturing (see further on for a discussion of these technologies).

Oil was first discovered in the Bakken Formation in 1951, and the formation has been the subject of a number of scientific studies since that time. Although experts agree that oil reserves in the formation are huge, they tend to disagree rather dramatically as to the fraction of oil and gas that can actually be recovered by any known technology. Those estimates range from a high of 50 percent (generally considered very optimistic) to a low of one percent, or even less. Recent reports by the USGS and the North Dakota Department of Mineral Resources tend to suggest that the lower rate of discovery may be more consistent with existing drilling technology. The actual amount of oil that can be produced from the Bakken formation depends on a number of factors, including the availability of new technology, the price of oil produced by conventional means, and other economic and environmental factors.

Thus far, some parts of the Bakken Formation have been found to be much more productive than others. The Elm Coulee oil field in Richland County, Montana, started producing oil in 2000 and, as of 2010, had produced 41 million barrels of oil and 24 billion cubic feet of natural gas. Production peaked at 100,000 barrels per day in 2006, and has since leveled off at about 70,000 barrels per day since that time. By 2007, the field had become the highest producing oil field in the onshore United States and accounted for more than half of all the oil being produced in the state of Montana. If other regions of the Bakken Formation are as productive as the Elm Coulee area, the region will be able to make a profound contribution to the nation's energy equation. Hopes that such a scenario will develop are, however, somewhat muted among experts familiar with the area ("More Oil?" 2012).

New Technologies: Horizontal Drilling and Hydraulic Fracturing

Most people probably think of the oil drilling process as driving a drill vertically into the earth until it reaches a stratum of rock containing oil and natural gas. The drill is then replaced by a pipe that allows the oil and gas to escape to the surface, where it can be captured and transported to a refinery. That scenario is largely accurate (if greatly oversimplified) for conventional oil wells. Indeed, early explorers for petroleum assumed that oil and gas deposits tend to be elliptical in shape, with the long dimension laid down vertically. That is, the oil and gas bed was thought to be very thick, but relatively narrow. As it turns out, that conception is not correct: oil and gas deposits tend to be elliptical in shape, but laid out in a horizontal direction. So, when a new well is drilled, it removes the oil and gas *directly* from only a small region of the overall reservoir and must wait while additional oil and gas flow into the well from either side of the drill hole.

An alternative method for drilling, however, is horizontal drilling (also known as directional drilling), in which a drill line is sent vertically into the earth until it reaches an oil and gas deposit. Its direction is then altered so that it enters the deposit along its main axis, pointing directly down the length of the deposit. A combination of sensors attached to the drill bit permits operators on the surface to determine the correct location at which the drill line should change direction. This approach tends to make it much easier to collect a much larger fraction of the resource than is possible with conventional vertical drilling.

The difference in drilling technologies is illustrated in the access a pipe has to a resource. The end of a vertical well is located within a reservoir that is typically 50–300 feet (15–100 meters) thick, so that oil must travel large horizontal distances to get to the pipe. Horizontal wells reach into the long end of a reservoir which, in some cases, may extend a distance of more than 5,000 feet (1,500 meters). The horizontal well can extract the resource directly through a much longer distance as drilling proceeds horizontally into the reservoir.

The first patent for a directional drilling technology was issued on September 8, 1891, to John Smalley Campbell, whose invention was originally designed for use in dental procedures. The patent indicated, however, that the same technology on a large scale could be used for other types of drilling, such as oil exploration. The first horizontal oil well was dug in 1929 by Robert E. Lee, of Coleman, Texas, at a well in Texon, Texas. Lee used the technology in an attempt to drain the oil out of a field, thus increasing its total output. Horizontal drilling was also explored as a recovery technology by other nations, including the Soviet Union, which drilled a horizontal well in Yarega, in northern Russia in 1937, and China, whose first research on horizontal technology dates to the mid-1960s, when one of two wells collapsed during drilling, and the second was abandoned during the nation's Cultural Revolution (Helms 2012).

When first introduced, the greatest drawbacks associated with horizontal drilling were the additional costs involved in locating the precise points at which drilling direction should be changed and then getting the drill line oriented in the proper direction. Advances in technology have largely resolved many of these issues, although the cost of horizontal drilling remains two to three times that of vertical drilling for a comparable environment. On the other hand, horizontal drilling can be so much more efficient than vertical drilling, that these higher costs are essentially ameliorated. It is not unusual for a horizontal well to produce 15 to 20 times the volume of oil and gas as a vertical well at the same site, making the higher drilling costs for the horizontal well relatively unimportant.

Horizontal wells have become increasingly popular because of the extended quantities of oil and gas they produce. The Independent Petroleum Association of America reported in 2010 that a turning point was reached in the U.S. petroleum industry in 2008 when, for the first time in history, more horizontal than vertical wells were drilled. In that year, the number of new horizontal rigs increased by 41 percent, whereas the number of conventional vertical rigs decreased by five percent. At that point, there were 8,078 horizontal wells in the United States, the largest number in Texas (4,214 wells), Oklahoma (1,056), North Dakota (675), and Arkansas (674) (*2009–2010 IPAA Oil and Gas Producing Industry* 2012).

A second process is commonly used in connection with horizontal drilling (and may also, but less commonly is, used with vertical drilling), that is, a process known as hydraulic fracturing or, more colloquially, *fracking* or *hydrofracking*. The use of this technology is often referred to as a frac job, in which a frac crew inserts a frac fluid or frac gel into a well using a frac pump or frac gun. (Terms using the expression *frac* may also be expressed as *frack*.) Fracking involves the injection under high pressure of a liquid mixture consisting primarily of water and sand (about 98% of the mixture), along with other chemicals into a well site. The force created by the injection process increases the pore size within the reservoir, allowing oil or gas to flow more readily toward the well pipe. Some of the five dozen or so chemicals most commonly used as additives in the fracking process and their functions are listed in Table 2.6. (The number of chemicals with potential use in the industry is actually much larger, close to 750 substances.)

Hydraulic fracturing is a relatively older technology in the oil and gas industry. A primitive form of the process was first used in the 1860s when nitroglycerin, in either solid or liquid form, was injected into a well and then ignited. The purpose of the process was to break up the rock formation holding the resource, making its extraction easier. The nitroglycerin process was both dangerous and, in some places, illegal, so industry researchers began to look for other means of achieving the same result. By the 1930s, acid injections were being used as a way of dissolving portions of the rocky material in which oil and gas were embedded, thus improving their flow. The first experiment using this more modern approach was conducted by Stanolind Oil in the Hugoton gas field in Grant County, Kansas, in 1947. The experiment did not result in any increase in oil or gas production, but it provided useful new data for the further development of the technology. Two years after the Hugoton experiment, the Halliburton Oil Well Cementing Company (Howco) put into operation the first commercial use of fracking technology at wells in Oklahoma and Texas, resulting in an

Table 2.6 Examples of Additives Used in Hydraulic Fracturing Technology

Chemical	Purpose in Fracking
Hydrochloric acid	Dissolves minerals and initiates cracks in rock
Quaternary ammonium chloride	Destroys bacteria that may corrode pipes
Ammonium persulfate	Prevents breakdown of fracking gel
Choline chloride	Stabilizes clay, preventing it from swelling or shifting
Sodium tetraborate	Maintains fluid viscosity
Ethylene glycol	Reduces friction within reservoir and pipes
Guar gum	Stabilizes sand–water suspension used in fracking
Citric acid	Prevents precipitation of metal oxides
Lauryl sulfate	Prevents formation of emulsions in the fracking fluid
Sodium carbonate	Maintains proper acidity in reservoir fluids
Sodium polycarboxylate	Prevents deposits of scale in pipes
2-Butoxyethanol	Acts as a surfactant on reservoir fluids

Source: "What Chemicals Are Used?" (2011).

increase of more than 75 percent in the oil and gas obtained from the treated wells. The technology appeared to have proved its value (Montgomery and Smith 2010, 27).

Use of the process soon began to grow by leaps and bounds. In the first year after Howco's initial use of fracking, 332 wells in the United States were treated with the technology. By the mid-1950s, about 3,000 new wells per month were being treated with fracking technology, and by 2008, more than 50,000 fracking operations had been completed worldwide. As of 2010, the vast majority of fracking operations on land and offshore were being conducted in the United States and Mexico, with 756 facilities in operation, followed by Canada (111), Latin America (72), Russia (49), China (32), non-China Asia (21), Africa (16), the Middle East (10), Europe (8), and Australasia (5). The total value of the fracking market is reported to have grown from less than $3 billion in 1999 to nearly $13 billion in 2007, with more than 80 percent of that market centered in the United States and Canada (Montgomery and Smith 2010, 27).

The use of hydraulic fracturing in oil and gas mining has become one of the most contentious energy issues in the United States in the early 21st century. Critics have raised a host of environmental and health concerns about the use of the technology, to which the oil and gas industry has responded aggressively. Both sides cite a large number of studies that support their specific stand on the use of fracking, resulting in an extended and heated debate that can be only briefly summarized in this book. For example, a report prepared by two scientists associated with TEDX, the Endocrine Disruption Exchange, found that 353 of the 632 chemicals used in natural gas fracking had one or more deleterious effects on human health, including "[m]ore than 75% [that] could affect the skin, eyes, and other sensory organs, and the respiratory and gastrointestinal systems. Approximately 40–50% [that] could affect the brain/nervous system, immune and cardiovascular systems, and the kidneys; 37% [that] could affect the endocrine system; and 25% [that] could cause cancer and mutations" (Colborn et al. 2012). A study by researchers at Duke University published in 2011 found another potential problem associated with fracking operations: the accumulation of methane gas in and around fracked wells. Methane concentrations at oil and gas fields in northeastern Pennsylvania and upstate New York where fracking was being used had average and maximum concentrations of 19.2 and 64 mg/L (milligrams per liter; 0.0192 and 0.064 oz/ft^3) of methane in drinking water, compared to an average level of 1.1 mg/L in water wells not associated with fracked wells. The Duke researchers pointed out that methane levels in regions undergoing fracking were sufficiently high to raise the question of possible explosions or fires resulting from the ignition of the methane (Osborn et al. 2011).

Concerns about fracking have been expressed in many parts of the world. Approval for use of the technology in new wells has been temporarily suspended in New South Wales, Australia, although it is still allowed in operating wells. The use of fracking is also under review for possible health and environmental effects in a number of places, including the Canadian provinces of Nova Scotia and Quebec and in the Karoo region of South Africa. In 2011, France became the first nation to adopt an outright ban on hydraulic fracturing because of potential environmental damage caused by the practice.

In the United States, regulation of fracking operations has been restricted largely to decisions by state and local agencies. In 2010, for example, the state of Pennsylvania imposed a ban on further drilling by the Cabot Oil & Gas Company until it sealed a well where fracking had been used, supposedly causing the contamination of drinking water in adjacent Dimock Township. In November 2010, the New York state senate voted to place a temporary moratorium on fracking until a number of health and environmental issues were resolved. And a number of towns, primarily in New York and Pennsylvania, but also in other parts of the nation, have adopted temporary or permanent bans on the use of hydraulic fracturing within their regions.

National regulatory policy on fracking in the United States, such as it is, is now defined by 2004 study conducted by the U.S. Environmental Protection Agency (EPA), which found that there was insufficient evidence that fracking produced significant health or environmental effects. That study was limited, however, to methane produced from coal beds and did not deal with the related problem of the effects of fracking in oil and gas exploration and mining. Because of increasing concerns about fracking during the first decade of the 21st century, a number of individuals and organizations have called for the EPA to revisit this question. In 2010, the U.S. House of Representatives Appropriation Conference Committee asked the EPA to investigate the potential harm to human health and the environment from fracking operations in the oil and gas industry. The EPA has now undertaken that study, with a report due by the end of 2012.

Quite naturally, the oil and gas industry has offered responses to each of the complaints raised about the potential risks of fracking in the recovery of fossil fuels. An example is the information provided on its website by the Keystone Energy Forum, an organization "committed to improving the public's understanding of, and support for, the many opportunities presented by the Marcellus Shale natural gas reserves here in Pennsylvania." The forum points, first of all, to the 2004 EPA study that concluded that "the injection of hydraulic fracturing fluids into CBM [coalbed methane] wells poses little or no threat to USDWs [underground drinking water sources] and does not justify additional study at this time." It also mentions a 1998 study by the Ground Water Protection Council of fracking operations in 25 states that found that "there was no evidence to support claims that public health is at risk as a result of the hydraulic fracturing of coalbeds used for the production of methane gas." (Note that both studies deal with coalbed methane, and not oil and gas obtained from fracked wells.) The forum also notes that huge benefits have accrued to consumers as the result of fracking. It argues that fracking will be required in 80 percent of all future oil and gas operations, and that such operations will result in "tens of thousands of additional jobs," making fracking an essential technology in the nation's energy future. Finally, the forum that the public has been protected in the past, and will continue to be protected, by industry best practices and by existing state and federal regulations of the use of fracking ("Hydraulic Fracturing Overview" 2012).

Energy Conservation

One approach for dealing with peak energy is an increased effort to conserve energy, that is, to use the energy resources currently available more efficiently so that they will last longer.

In his remarkable essay on the history of energy conservation, "The Modern History of Energy Conservation: An Overview for Information Professionals," Donald R. Wulfinghoff, publisher of the Energy Institute Press, divides the topic into two general sections: the early history of energy conservation and the modern history of energy conservation. The first of these periods was characterized, Wulfinghoff explains, by the development of new forms of energy that allowed humans to accomplish more work with lesser input of human energy. During this period, the greatest amount of energy input shifted from human efforts to the use of animals to the development of machines to the invention of a variety of forms of energy, such as wind, solar, geothermal, tidal, and fossil fuels. The modern era of energy conservation, Wulfinghoff says, dates from the early 1970s, when the Arab oil embargo dramatically reduced the availability of petroleum and natural gas to the developed world, especially to the United States, which was most seriously affected by the embargo. The shock of that action caused nations to begin to think of the risks of running out of abundant, inexpensive energy for the first time, making energy conservation a serious national social, economic, and political issue. While that shock has affected nations around the world, it has, for the past half century, been most noticeable in the United States, the world's largest consumer of energy on a per-person basis by far (Wulfinghoff 2012).

One of the first responses to the Arab oil embargo of 1973 by the U.S. government was the passage of laws mandating increased energy conservation and efficiency and the creation of a variety of governmental agencies responsible for carrying out these mandates. For example, the Energy Policy and Conservation Act (Public Law 94-163) dealt with a number of issues related to the domestic production and consumption of energy. Among the many provisions of the act was a requirement that certain types of power plants burn coal rather than oil and gas, that loan guarantees be provided for the development of new coal mines, that the president have power to restrict the exportation of fossil fuels, that the president have the authority to allocate fossil fuel resources on the basis of national priorities, that a Strategic Petroleum Reserve be established, and that an Early Storage Reserve be developed for certain types of fuels. The act also established an energy conservation program and standards for many different types of industrial and home appliances, standards that were later modified and extended by the National Appliance Energy Conservation Act of 1987 (Public Law 100-357), the Energy Policy Act of 1992 (Public Law 102-486), and the Energy Policy Act of 2005 (Public Law 109-58).

Energy conservation achieved formal administrative recognition in 1971 when President Richard M. Nixon created the Office of Energy Conservation within the Department of the Interior. That office was not particularly active or effective until it was reinvigorated under President Jimmy Carter in 1975 and the Energy Research and Development Administration (ERDA) within the Department of Energy, charged with expanding energy technologies available to U.S. citizens. Energy conservation programs again received something of a setback under President Ronald Reagan, who redesigned ERDA with a more limited function under the name of the Office of Conservation and Solar Energy. Most recently, that office has once more been given a revised mandate and renamed the Office of Energy Efficiency and Renewable Energy.

The flurry of attention to energy conservation in the 1970s slowed significantly during the 1980s, as the federal government gradually lost interest in the topic and restricted itself to the expansion of existing laws. During both the 1970s and the 1980s, energy conservation was seen largely as an issue of supply and demand, with governmental efforts aimed at reducing the demand for energy among the American public and encouraging individuals and companies to use energy more efficiently. By the 1990s, however, that scenario had begun to change, as governmental bodies and the general public began to see the connection between energy use and environmental issues, most especially the problem of global climate change. Much of the emphasis within energy conservation programs since the 1990s has had that environmental concern as a major component of conservation and efficiency programs.

The future of energy conservation in the future energy equation for the United States is uncertain. There currently seems to be relatively little appetite among federal and most state legislatures for developing and imposing further conservation standards, as occurred during the 1970s. As only one example, the U.S. government first established fuel efficiency standards (Corporate Average Fuel Economy, or CAFE, standards) for passenger vehicles in 1978. The original standard, which new automobiles were required to meet, was 18.0 mpg (miles per gallon). That standard was raised every year until 1985, when it reached 27.5 mpg. It was then lowered for a period of years until 1990, when it was returned to 27.5 mpg. It remained at that level for the next 20 years, when it was once more raised to 30.2 mpg. The fact that the standard remained at 27.5 mpg for such an extended period of time suggests that a series of both Democratic and Republican U.S. presidential administrations did not see fit to require automobile companies to increase the efficiency of their products. In July 2011, President Barack Obama announced that he was raising the CAFE standard to 54.5 mpg as of 2025, although it is somewhat difficult to know for sure if that decision will survive nearly 15 years into the future (Klier and Linn 2012).

Efforts at energy conservation are hardly unique to the United States. Many developed countries in the world have experienced crises in energy availability and consumption like those in the United States, with a variety of efforts to deal with possible energy shortfalls and increasing environmental impacts. China is one such country that appears to be repeating some of the experiences of the United States over the past half century. Prior to 1980, the Chinese government followed a Soviet-style energy policy, with emphasis on the expansion of industry with the concomitant expense of energy resources at the fastest possible rate. The concept of limiting the use of energy was unknown to the country during that period. As a result, the Chinese had the most energy-intensive and energy-inefficient system in the world between 1949 and 1980.

A dramatic change occurred in 1980, however, with the ascension of Deng Xiaoping to the country's place of leadership. In a series of meetings, a group of Chinese academics convinced the new premier that China was facing an energy crisis and that governmental controls were needed to bring energy consumption and production under control. Deng responded favorably to these suggestions, and a number of reforms were instituted, including controls on the consumption of energy by factories, with close monitoring of those regulations; the closing down of inefficient facilities; implementation of controls on the use of certain types of fossil fuels; reductions in taxes on the most efficient operations; financial support for the development of alternative technologies; and education and training for corporate managers and the general public on the efficient use of energy.

For almost two decades, these reforms appeared to be effective, bringing China's energy use under control. But the conversion of the nation to a market economy in the late 1990s changed that trend. As the economy began to boom under market conditions, many of the 1980s reforms were abandoned or ignored. A new tax code adopted in 1994, for example, eliminated the tax incentives for efficient factory operations. As a consequence, the economy as a whole and the Chinese energy industry in particular once more began to grow out of control. As the nation rapidly moves toward becoming the dominant country in the world, it is unclear what role, if any, energy conservation and efficiency will have in the nation's economic, political, social, and military future. As one group of experts who have studied this problem have written, the key question for the near future is how China can reduce the growing demand for energy and still meet all its energy needs? "Unless this problem is solved," these experts say, "China's economic goals will be placed in jeopardy and the environmental consequences of energy policy failure are truly frightening" (Levine 2012).

Environmental Impacts of Fossil Fuel Use

The combustion of fuels has been associated with environmental problems for at least two millennia. In a typical very old record, the Roman philosopher Seneca complained in 61 CE about the polluted air in Rome caused by the burning of wood and coal. "As soon as I had gotten out of the heavy air of Rome and from the stink of the smokey chimneys thereof," he once wrote, "I felt an alternation of my disposition" (quoted in Vallero 2005, 109). Many of the earliest reports of air pollution resulting from the burning of fuels come from Great Britain. In 1157, for example, Queen Eleanor demanded permission to move out of Tutbury Castle in Nottingham because air pollution from the burning of wood had become unendurable. Laws attempting to limit such pollution appeared early in the country. In 1306, for example, King Edward I prohibited the use of coal for fires, declaring that "Whosoever shall be found guilty of burning coal shall suffer the loss of his head" (Newton 2007, 3).

The rise of the Industrial Revolution made such problems far more extensive as nations moved to the combustion of coal to drive the many new inventions developed during the period. Stories and illustrations from the period emphasize the smokey conditions that dominated any English city with any type of industrial operation (which was essentially all of them). The government did not take much official notice of these problems until 1819 when Parliament appointed a committee to study the problems associated with air pollution from the burning of coal. Still, it took almost three decades before such studies produced the first legislation relating to polluted air, the Public Health Act of 1848. The act called for a newly formed agency, the Health Agency, to control the release of smoke and ash in English cities. The government's attitude about such efforts is reflected in the fact, however, that it took another 24 years before the first director of that office was appointed, Robert Angus Smith, in 1872. Charles Dickens provided one of the most vivid descriptions of the situation in English cities of the time in a passage from his *Bleak House.* He wrote that "[s]moke lowering down from chimney-pots, making a soft black drizzle, with flakes of soot in it as big as full-grown snow flakes—gone into mourning, one might imagine, for the death of the sun" (Dickens 1853, 1).

The Industrial Revolution resulted not only in unprecedented industrial and economic progress, therefore, but also a rise in environmental disasters caused by the combustion of fossil fuels. London has had an especially notable history of events caused by a combination of smoke produced by the combustion of (primarily) smoke and the city's notorious foggy weather, a condition that was eventually given the name of *smog* (*smoke* + *fog*). The first such event in which deaths were reported occurred in 1873, when more than 500 people were killed by a variety of respiratory conditions caused by breathing polluted air. Similar events occurred again in 1880 (with a toll of more than 2,000 deaths), 1892 (more than 1,000 deaths), 1952 (the worst case in recorded history, with more than 4,000 deaths), and 1962 (750 deaths). Smog disasters occurred in other parts of the world also, notably in Meuse, Belgium, in 1930 (63 deaths); Donora, Pennsylvania, in 1948 (20 deaths); Pozo Rica, Mexico, in 1950 (22 deaths); and New York City in 1953 (170–260 deaths). Other smog-related events were also being reported in cities around the world that did not result in large number of deaths, but did cause a reduction in visibility and severe respiratory problems for residents ("Major Air Pollution Disasters" 2012).

One might well ask why nations tolerated disasters such as these from pollution caused by fossil fuel use. One important reason was that environmental degradation was historically accepted as a sign of success. Severe pollution was an indication of a prosperous economy. The more successful industrial operations were, the higher the standard of living for at least some of the nation's population—certainly the upper class, who owned the factories, and often the new and growing middle class. If successful factories also released excessive amounts of hazardous waste products into rivers and lakes and into the air, that was perhaps unfortunate, especially for members of the working class, but it seemed to be an unpreventable by-product of a nation's overall economic success. (That viewpoint remains a powerful argument in the 2010s when many politicians object to greater environmental controls on industrial activity because it may tend to retard economic progress.)

In fact, it was not until the second half of the 20th century that most nations around the world began to institute controls on the use of fossil fuels to reduce pollution. One of the first of these laws was the Public Health (Smoke Abatement) Act adopted by the British Parliament in 1926. Like much of the legislation that had preceded it, the act was relatively weak and poorly enforced. The earliest federal legislation in the United States included the 1948 Federal Water Pollution Control Act and the 1955 National Air Pollution Control Act, both similar to British legislation in being weak and poorly enforced. Stronger legislation in the United States was not adopted for another two decades with passage of the National Environmental Policy Act (1969), the Clean Air Act (1963; amended in 1965, 1966, 1967, 1969, 1970, 1977, and 1990), the Federal Water Pollution Control Act (1972; amended in 1977 and 1987), the Occupational Safety and Health Act (1970), and a host of similar legislation dealing with environmental issues.

Common Air Pollutants

The EPA now tracks the level of air pollution in a variety of ways, most commonly through the use of six so-called *criteria pollutants,* pollutants that are found commonly throughout the United States. These criteria pollutants are carbon monoxide, oxides of nitrogen, sulfur dioxide, particulate matter (PM), lead, and ozone. These pollutants are produced from many different sources, most of them related to the combustion of fossil fuels. For example, carbon monoxide is produced (along with carbon dioxide and water) when any fossil fuel burns. Oxides of nitrogen and sulfur dioxide are also produced during the combustion of fossil fuels because nitrogen and sulfur compounds always occur as trace contaminants in all kinds of fossil fuel, although more abundantly in some forms of coal, oil, and natural gas than in other forms. PM is another name for very finely divided solid particles that are part of the ash that forms when a fossil fuel burns. Various types of PM are usually differentiated from each other on the basis of the size of particles of which the pollutant exists. For example, particulates designated as $PM_{2.5}$ consist of particles less than 2.5 micrometers in diameter, whereas PM_{10} particulates consist of particles less than 10 micrometers in diameter. Authorities make this distinction between the health effects of particulates differs to some degree depending on the size of the particle that enters the body.

The EPA in the United States, the European Environment Agency, the World Health Organization (WHO), and other groups have been collecting data about the levels, sources, and health effects of these six criteria pollutants (and other air pollutants) over many years, dating at least to the 1970s. Table 2.7 shows the trends for these pollutants in the United States over the past four decades. Trends in ozone concentrations are well represented by trends in the concentration of volatile organic compounds which, when exposed to sunlight, are a major source of ozone production in the atmosphere.

Concerns about air pollutants are based largely on their potential effects on the health of humans and other animals and plants. These health effects have now been studied in extensive detail for many years and are relatively well known. They depend largely on three factors: the nature of the pollutant itself, the concentration to which a plant or an animal is exposed, and the length of time over which the organism is exposed to

Table 2.7 Trends in Pollutants in the United States, 1970–2010 (Thousands of Tons)

Carbon Monoxide		
Source	**1970**	**1980**
Electricity generation	237	322
Industrial operations	770	750
Other fuel combustion	3,625	6,230
Metal processing	3,644	2,246
Petroleum and related industry	2,179	1,723
Waste disposal and recycling	7,059	2,300
Highway vehicles	163,231	143,827
Off-road vehicles	11,371	16,685
Total	204,403	185,407
Fuel combustion	4,632	7,302
Industrial processes	16,899	9,250
Transportation	174,602	160,512
Oxides of Nitrogen		
Source	**1970**	**1980**
Electricity generation	4,900	7,024
Industrial operations	4,325	3,555
Other fuel combustion	836	741
Metal processing	77	65
Petroleum and related industry	240	72
Waste disposal and recycling	440	111
Highway vehicles	12,624	11,493
Off-road vehicles	2,652	3,353
Total	26,883	27,079
Fuel combustion	10,061	11,320
Industrial processes	1,215	666
Transportation	15,276	14,846
Sulfur Dioxide		
Source	**1970**	**1980**
Electricity generation	17,398	17,469
Industrial operations	4,568	1,842
Other fuel combustion	1,490	971
Metal processing	4,775	1,842
Petroleum and related industry	881	734
Waste disposal and recycling	8	33

1990	1995	2000	2005	2010
363	372	484	642	728
879	1,056	1,219	1,150	1,015
4,269	4,506	3,081	3,330	2,926
2,640	2,380	1,295	829	803
333	348	161	350	343
1,079	1,185	1,849	1,554	1,565
110,255	83,881	68,061	48,544	35,605
21,447	23,874	24,178	20,672	9,706
154,186	126,777	114,467	93,030	67,790
5,511	5,934	4,784	5,123	4,670
5,853	5,791	4,479	3,585	3,429
131,702	107,755	92,239	69,216	45,312

1990	1995	2000	2005	2010
6,663	6,384	5,330	3,598	2,096
3,035	3,144	2,723	1,812	1,547
1,196	1,298	766	730	656
97	98	89	66	72
153	110	122	249	462
91	99	129	146	125
9,592	8,876	8,394	6,492	4,283
3,781	4,113	4,167	4,887	2,873
25,529	24,956	22,598	18,914	12,914
10,894	10,826	8,819	6,141	4,299
891	873	943	1,124	1,136
13,373	12,989	12,560	11,379	7,157

1990	1995	2000	2005	2010
15,909	12,080	11,396	10,156	5,144
3,550	3,357	2,139	1,735	1,2909
831	793	628	582	433
726	530	313	175	162
430	369	316	199	143
42	47	34	28	27

(continued)

Table 2.7 (*continued*)

Source	1970	1980
Highway vehicles	273	394
Off-road vehicles	278	323
Total	31,218	25,925
Fuel combustion	23,456	21,391
Industrial processes	7,101	3,807
Transportation	551	717
Particulate Matter (PM_{10})		
Source	**1970**	**1980**
Electricity generation	1,775	891
Industrial operations	641	679
Other fuel combustion	455	887
Metal processing	1,316	622
Petroleum and related industry	286	138
Waste disposal and recycling	999	273
Highway vehicles	480	432
Off-road vehicles	164	257
Total	13,023	7,013
Volatile Organic Chemicals		
Source	**1970**	**1980**
Electricity generation	30	45
Industrial operations	150	157
Other fuel combustion	541	848
Metal processing	394	273
Petroleum and related industry	1,194	1,440
Waste disposal and recycling	1,984	758
Highway vehicles	16,910	13,869
Off-road vehicles	1,616	2,192
Total	34,659	31,106
Fuel combustion	721	1,050
Industrial processes	14,311	12,862
Transportation	18,526	16,061

Source: U.S. Environmental Protection Agency (2012).

1990	1995	2000	2005	2010
503	335	260	146	35
371	406	437	834	165
23,076	18,619	16,347	14,594	7,938
20,290	16,230	14,163	12,473	6,866
1,901	1,638	1,418	1,016	794
874	741	697	980	200

1990	1995	2000	2005	2010
295	268	687	627	429
270	302	320	350	207
631	610	465	464	383
214	212	140	83	69
55	40	38	28	25
271	287	362	289	292
387	304	230	193	117
328	339	322	345	170
27,752	25,819	23,747	21,149	10,778

1990	1995	2000	2005	2010
47	44	62	48	46
182	206	173	132	111
776	823	949	589	219
122	125	67	49	43
611	642	428	562	1,232
986	1,067	415	392	181
9,388	6,749	5,325	4,112	3,147
2,662	2,890	2,644	2,866	1,313
24,108	22,041	17,512	18,421	13,443
1,005	1,073	1,184	769	376
9,994	10,779	7,626	7,382	5,292
12,050	9,639	7,969	6,978	4,460

the pollutant. For example, exposure to low concentrations of carbon monoxide for brief periods of time may cause headaches, disorientation, nausea, and similar modest neurological and respiratory effects. Exposure to larger concentrations may result in more serious consequences, such as loss of consciousness, coma, and death.

Many national, international, occupational, and other organizations have established safe exposure limits for a variety of air pollutants. Table 2.8 shows the most recent recommendations of this type from the European Commission. These recommendations tend to be similar to, but not always identical to, those of other agencies, such as the Canadian Air Resources Board, the U.S. National Institute for Occupational Safety and Health, and the WHO. They usually indicated the concentration of a pollutant to which one can be exposed and the time over which exposure should occur.

Acid Precipitation

Human health is by no means the only environmental concern associated with the combustion of fossil fuels. Researchers have known for decades, for example, that air pollutants can have harmful effects on plants as well as on animals. Oxides

Table 2.8 Recommended Exposure Limits for Certain Air Pollutants (European Commission)

Pollutant	Exposure Concentration	Exposure Time
Carbon monoxide	10 mg/m^3	8 hours
Nitrogen dioxide	200 µg/m^3	1 hour
	40 µg/m^3	1 year
Sulfur dioxide	350 µg/m^3	1 hour
	125 µg/m^3	24 hours
Particulate matter: $PM_{2.5}$	25 µg/m^3	1 year
Particulate matter: PM_{10}	50 µg/m^3	24 hours
	40 µg/m^3	1 year
Ozone	125 µg/m^3	8 hours

Source: European Commission (2011).

of sulfur and nitrogen can enter the stomata of leaves and give rise to sulfate, sulfite, nitrate, and nitrite ions that are toxic to the plant. Ozone is a strong oxidizing agent that can attack and destroy leaves, stems, flowers, and other plant parts. PM can collect on leaves and block stomata, lowering their access to carbon dioxide and shield the plant from sunlight, reducing the rate of photosynthesis. On a much larger scale, fossil fuel emissions can have widespread impact on large areas of plant life, as is the case with acid precipitation (also known as acid deposition or simply as acid rain).

Acid precipitation occurs when fossil fuels containing sulfur and, less commonly, nitrogen as impurities are burned in electricity-generating plants and other industrial factories. The combustion of these two elements results in the formation of sulfur dioxide and nitrogen oxides during combustion. These oxides then combine with oxygen and moisture in the air to produce sulfuric acid (sulfate) and nitric acid (nitrate):

$$S + O_2 \rightarrow SO_2 + (O) + H_2O \rightarrow H_2SO_4$$

$$N + O_2 \rightarrow NOx + (O) + H_2O \rightarrow HNO_3$$

When that moisture returns to the Earth's surface in the form of rain, snow, sleet, hail, or some other form of precipitation, sulfuric and nitric acid settle on plant leaves or drop to the ground, where they become part of the groundwater system and are taken up by trees, grass, crops, and other plants, causing them damage or killing them. Acid precipitation also falls on bodies of water, decreasing their pH (increasing their acidity), sickening or killing plants and animals that live in the water.

Acid precipitation was first observed as far back as the mid-20th century when scientists attributed the die-off of extensive regions of forest in Sweden, Norway, Germany, Japan, and eastern Canada and the United States to damage by acid precipitation. They also found that the water in lakes and ponds in these areas was much more acidic than normal, resulting in large-scale die-offs of fish and other aquatic organisms. Researchers suggested that prevailing west-to-east winds were blowing clouds of acidic moisture from areas of high industrial output, such as England, the Ruhr area of Germany, and the American Midwest, into the regions experiencing damage.

By the end of the 20th century, a number of nations whose industries were supposedly responsible for acid precipitation damage had instituted programs to reduce fossil fuel emissions. In the United States, the primary program was the Acid Rain Program, created as Title IV of the Clean Air Act of 1990. That act specified a number of steps that had to be taken to reduce the amount of acid rain pollutants produced by American industries. The program is widely praised as being one of the most effective pieces of legislation for dealing with an environmental issue related to the use of fossil fuels. Studies show that the amount of sulfate and nitrate deposition in the various parts of the United States and Canada decreased by as much as 60 percent between 1989 and 2008, and by an average of 44 percent in the Midwest, 38 percent in the mid-Atlantic, and 43 percent in the Northeast. Other nations have also adopted legislation similar to the Acid Rain Program, most famously of which was the so-called Sulfur Treaty, formally the Protocol to the 1979 Convention on Long-Range Transboundary Air Pollution on the Reduction of Sulphur Emissions or Their Transboundary Fluxes by at Least 30%, which entered into force in 1987, and its amended version, Protocol to the 1979 Convention on Long-Range Transboundary Air Pollution on Further Reduction of Sulphur Emissions, which entered into force in 1998. The latter treaty has 24 signatory nations, Austria, Belgium, Bulgaria, Canada, Croatia, Czech Republic, Denmark, Finland, France, Germany, Greece, Hungary, Ireland, Italy, Liechtenstein, Luxembourg, Netherlands, Norway, Slovakia, Slovenia, Spain, Sweden, Switzerland, and United Kingdom. Studies of the effectiveness of the Sulfur Treaty in reducing fossil fuel emissions seem to be encouraging. By 1993, emissions from most signatory nations had decreased substantially over

1980 levels: as much as 82 percent in Austria, 80 percent in Sweden, and 79 percent in Finland, for an overall improvement of 43 percent among all signatory countries. Only two countries showed no improvement at all, Greece (an increase of 36% in emissions) and Portugal (13%). As with the U.S. Acid Rain Program, the European Sulfur Treaty appears to have been a remarkable success in resolving the issue of environmental damage caused by acid precipitation (Sulfur Treaty 2012).

Clean Air Legislation

Many nations have also enacted laws dealing with the more general problem of air pollution caused by fossil fuel combustion. In the United States, the core legislation on air pollution is to be found in the Clean Air Act of 1970 and its amended versions, the Clean Air Act of 1977 and the Clean Air Act of 1990. The EPA has projected that the latest of these acts will result in preventing 230,000 early deaths by the year 2020. Many other nations around the world have adopted legislation similar to the Clean Air Act and its amendments. Some examples are the Ley General del Ambiente (General Environmental Law) of 2002 (Argentina), Cleaner Production Promotion Law of 2002 and Raising Taxes for Pollutant Discharge law of 2007 (China), Rational Use of Energy Law of 1996 (Costa Rica), Air (Prevention and Control of Pollution) Acts of 1981 and 1988 (India), Air Pollution Abatement Programme of 1998 (Iran), Environmental Quality (Clean Air) Regulations of 1978 (Malaysia), Environmental Impact Assessment (EIA) Decree No. 86 of 1992 (Nigeria), Air Quality Act of 2004 (South Africa), and the Control of Air Pollution from Heating/Motor Vehicles/Industrial Plants of 2004/2005/2006 (Turkey). Legislation of this kind is almost certainly related to the reduction of harmful emissions from fossil fuel combustion, as suggested for the United States in Table 2.7 (SD-PAMS Data Base 2012).

The conflict between aggressive environmental legislation designed to reduce fossil fuel emissions and possible deleterious effects on a nation's economy through the enforcement of such legislation is often a theme of current events in many countries and states. As an example, President Barack Obama announced in September 2011 that he was abandoning the adoption of newer and more stringent ozone emission rules proposed by the EPA. Those rules would have reduced recommended emission levels for ozone from 75 parts per billion to either 60 or 70 parts per billion. That change would have put hundreds of counties across the United States out of compliance and would have required many industries to make costly changes in their operations. Business leaders complained that the new regulations would have cost industries millions of dollars and thousands of jobs at a time of severe economic distress in the country. Obama was convinced by the argument that economic factors trumped environmental costs in these cases, and the EPA's new ozone standards were rejected.

Global Climate Change

By the 1980s, the world had begun to recognize yet another potential by-product of the extensive use of fossil fuels: global climate change. Data began to accumulate suggesting that the combustion of fossil fuels over the preceding two centuries had released so much carbon dioxide into the atmosphere that the planet's climate was beginning to change, resulting in a slow but steady increase in the annual average global temperature. Probably the most famous data set of this kind ever collected was the result of research by American geochemist Charles David Keeling, who measured carbon dioxide concentrations in the atmosphere above Mauna Loa, Hawaii, over a period of 50 years. Keeling found that the concentration of carbon dioxide increased slowly but surely throughout that period of time, as shown in Figure 2.1.

Data like those provided in the Keeling curve have provided a scenario to which most atmospheric scientists today subscribe, namely that the concentration of carbon dioxide in the Earth's

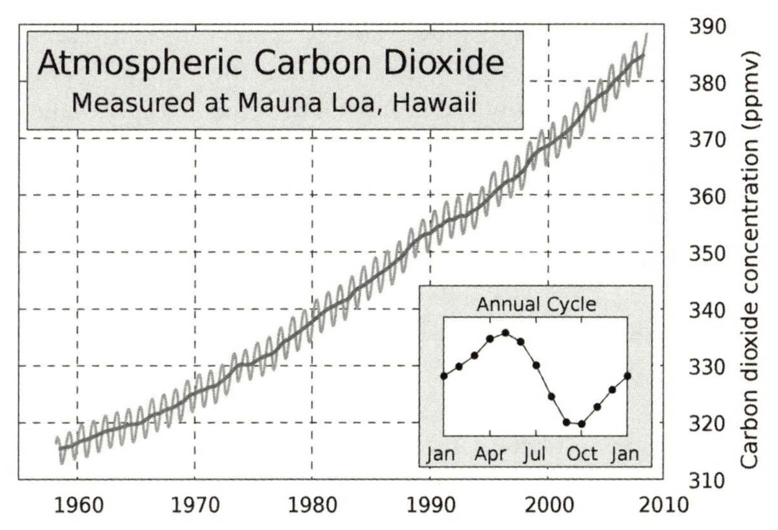

Figure 2.1 Keeling curve. National Oceanic and Atmospheric Administration, Earth System Research Laboratory, Global Monitoring Division.

atmosphere is increasing (there is not much debate about this point), which is likely to result in a corresponding increase in the planet's annual average temperature (again, not a lot of disagreement), which is likely to result in some significant long-term changes in climate (somewhat less agreement about possible trends), with significant effects on weather patterns, agricultural practices, disease patterns, and a number of other factors in the natural environment and human civilization (quite a bit of disagreement about the details of these changes).

Despite some disagreement about the changes that are likely to accompany increasing levels of carbon dioxide in the atmosphere, scientists and politicians have become increasingly concerned about the ever-growing release of carbon dioxide into the atmosphere as the result of fossil fuel burning. In response to these concerns, representatives from 196 nations met in Kyoto, Japan, in December 1997 to find ways to reduce the emission of greenhouse gases that tend to contribute to global warming. The result of that meeting was adoption of the Kyoto Protocol, an amendment to the United Nations Framework Convention on Climate Change, originally adopted at the United Nations Conference on Environment and Development (the so-called Earth Summit) held in Rio de Janeiro in June 1992. The Kyoto Protocol established a complex set of benchmarks to which signatories of the protocol agree, reducing greenhouse gas emissions to earlier levels by some specified date in the future. Of the 196 attendees, 191 states eventually signed and ratified the treaty; one signed, but has not ratified the treaty (the United States); and three other nations have neither signed nor ratified the treaty: Afghanistan, Andorra, and the Vatican City.

The position of the United States vis-à-vis the Kyoto Protocol has long been that it would not ratify any treaty that established limits on greenhouse gas emissions for developed nations that were different from those for developing nations. In fact, the U.S. Senate passed unanimously a "sense of the Senate" resolution on July 25, 1997, establishing this policy. When the Kyoto Protocol was passed with just this kind of provision, it seemed to be essentially dead-on-arrival in the U.S. Senate, which, in fact, was the case. Instead of agreeing to the protocol provisions, the United States decided to go it alone on climate change legislation. In 2002, President George W. Bush announced the nation's policy on the issue, a policy that remains largely in effect in 2012. The policy is that the federal government should not impose limitations on industrial activities to reduce greenhouse gas emissions because such actions will have deleterious effects on the national economy. Instead, industry should be encouraged to adopt such limitations as they feel will be environmentally beneficial, but that will not interfere with their continuing growth and development.

As might be expected, various nations have had varying degrees of success in meeting the targets for the emission of greenhouse gases under the provisions of the Kyoto Protocol. A February 2011 Fact Sheet issued by the United Nations Framework Convention on Climate Change shows that some countries continue to emit greenhouse gases far in excess of the target set by the treaty, such as Australia (an increase in greenhouse gas emission of 35.0% between 1990 and 2010), Canada (+38.2%), Greece (+37.5%), Iceland (+37.7%), Portugal (+46.7%), Spain (+52.5%), and Turkey (+157.6%). Countries that have shown a decrease in greenhouse gas emissions during this period included Estonia (–56.0%), Hungary (–28.5%), Latvia (–46.1%), Lithuania (–39.9%), and Ukraine (–47.9%) ("Fact Sheet" 2012).

Renewable Energy

Any discussion about the world energy crisis and/or energy peaks inevitably involves some consideration of renewable energy resources. The fundamental issue associated with the use of fossil fuels is that such fuels are generally regarded as limited; most experts agree that they were produced at one time in the Earth's history under a very special set of circumstances that have not been and are not likely to be replicated at any time in the future. Thus, the most basic question of all with regard to energy use is what humans will do when fossil fuel resources are exhausted, whether that happens in 10 years, a hundred years, or a thousand years from now. Possible alternative renewable sources of energy that are usually mentioned include nuclear power, wind energy, solar power, geothermal energy, biomass, and various forms of ocean energy.

Nuclear Power

The discovery of nuclear fission and nuclear fusion in the 1930s and 1940s raised the hope of many scientists and industrialists that these sources would eventually provide dependable, cheap, and safe sources of power that could eventually replace the world's dependence on fossil fuels. The first nuclear power plant in the world was the Calder Hill nuclear power station, at Sellafield, England, with an initial capacity of 50 MW (megawatts), later increased to 200 MW. The Calder Hill plant began operations on August 27, 1956, and was finally decommissioned on March 31, 2003. Interest in nuclear power grew rapidly at first, with 82 plants online as of 1970, 168 plants as of 1975, 244 plants as of 1980, 366 plants as of 1985, and 420 plants as of 1990. Plant construction then began to level off and has remained relatively constant ever since, with 436 plants in 1995, 441 plants in 2000, 443 plants in 2005, and 433 plants in 2011 ("Nuclear Power Reactors in the World" 2012, Table 8, 18–19).

Diminished interest in nuclear power was prompted toward the end of the 20th century by two factors, the first of which was two disasters at nuclear plants, one at the Three Mile Island Nuclear Power Plant in Dauphin County, Pennsylvania, on March 28, 1979, and the other at the Chernobyl Nuclear Power Plant in Ukraine on April 26, 1986. Both disasters highlighted the potential risks associated with nuclear energy, risks that had largely been ignored or discounted in the decades before they occurred. The second factor affecting the construction of new nuclear power plants was the unanswered question as to how wastes from nuclear power plants would be disposed of, a question that has yet to be satisfactorily answered anywhere in the world. In the United States, for example, a plan to store the nation's nuclear wastes in a specially built depository in Yucca Mountain, Nevada, originally adopted in 1987, was finally abandoned in 2009 largely as a consequence of ongoing objections raised by the state of Nevada. As a result, the United States still has no permanent depository for high-level nuclear wastes that have been accumulating since the mid-1940s.

Nuclear power is very far from being dead as a possible option for the world's future energy needs, however. A number of nations rely to a very large extent on nuclear power for meeting their own national needs, including France, which obtains 74.1 percent of all of its power from nuclear sources, Slovakia (51.8%), Belgium (51.1%), Ukraine (48.1%), Hungary (42.0%), Armenia (39.4%), Sweden (38.1%), Switzerland (38.0%), Slovenia (37.4%), the Czech Republic (33.2%), Bulgaria (33.0%), South Korea (32.2%), and Japan (29.2%).

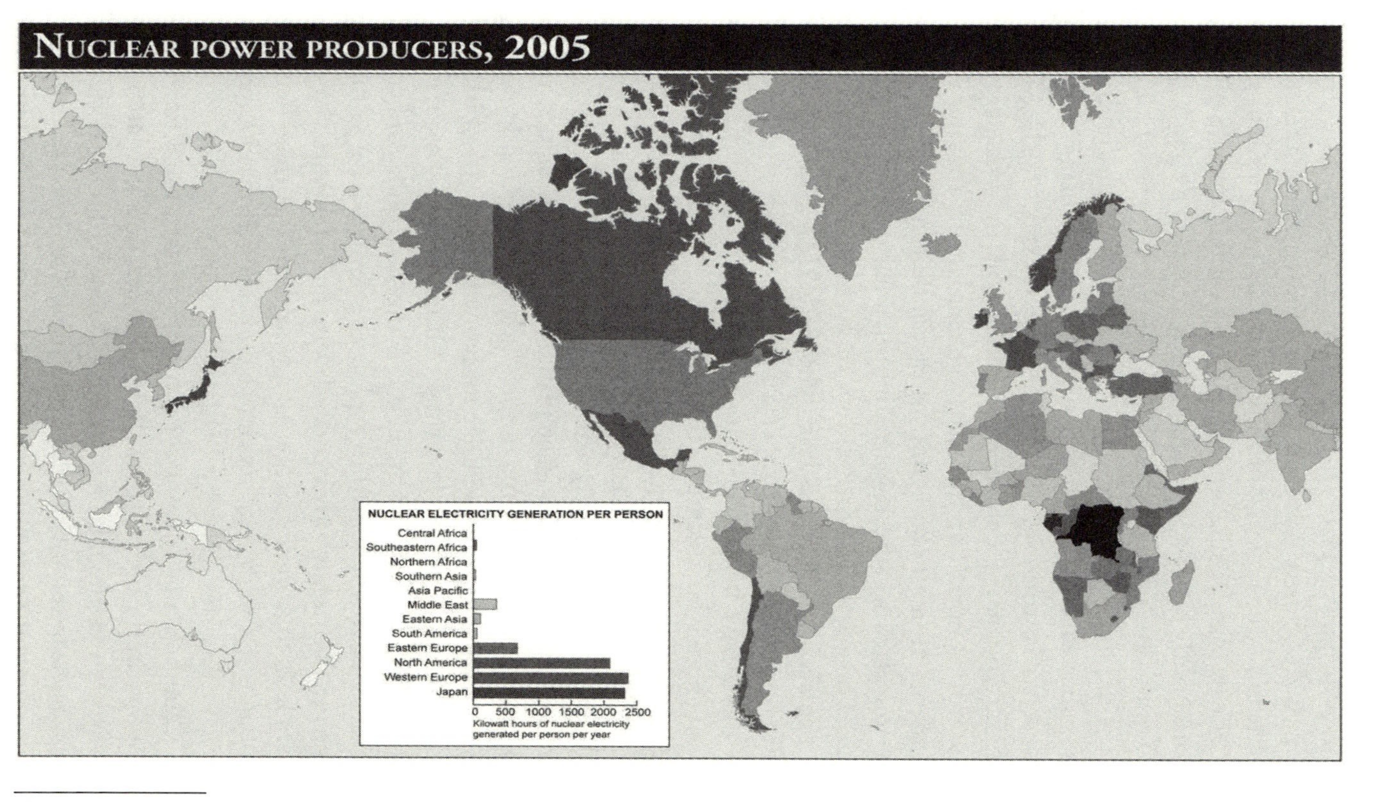

Nuclear power producers and use, 2005. (ABC-CLIO)

Nuclear power plant construction has also begun to pick up, with 65 new plants in 14 countries currently under construction. The majority of those plants are being built in China (27 new plants), followed by Russia (11 plants), India (8 plants), and South Korea (5 plants) ("Nuclear Power Reactors in the World" 2012, Table 8, 18–19).

The push for nuclear power as a substitute for diminishing supplies of coal, oil, and natural gas is also being seen in a part of the world where it might be least expected, the Middle East. Apparently convinced that peak oil and peak gas have already occurred, or are likely to occur within a matter of decades, a number of oil- and gas-rich nations have begun to develop plans for the construction of nuclear power plants on their own soil. The first nation in the Middle East to actually open such a facility was Iran, whose 450-MW plant at Busher began operation in October 2011. A number of other Arab nations have begun negotiations with a variety of oil and gas companies. For example, the U.S. government has signed a nuclear cooperation agreement with the United Arab Emirates and Tunisia; France has signed similar pacts with Kuwait and Tunisia; Russia not only was instrumental in development and construction of the Busher plant, but also has agreed to aid Kuwait in development of its first nuclear reactor; and Jordan has signed an agreement with Japan that will allow Mitsubishi and Toshiba to build reactors there. Other Middle East nations, including Algeria, Egypt, Libya, Morocco, Saudi Arabia, and Sudan, have just started to explore the development of nuclear power. Reflecting this trend, Abdelmajid Mahjoub, chairman of the Arab Atomic Energy Agency, said in 2011 that "[t]he use of atomic energy is an inevitable choice in the development of Arab countries" (Newby 2012).

The potential for nuclear power to replace fossil fuel technology in the near future suffered a severe blow in March 2011 when an earthquake and tsunami severely damaged three of six nuclear reactors at the Fukushima Dai-ichi nuclear facility on Japan's eastern coast. (Two other reactors had been shut down for maintenance, and a third was shut down while being defueled.) Although fewer than five deaths have been directly attributed to events at the power plant itself, 300,000 people were relocated because of possible exposure to radiation, and the Tokyo Electric Power Company that maintains the plant expects to see losses of more than $100 billion. To some observers, the disaster marked the end of the nuclear era. A report issued by the Worldwatch Institute in April 2011 concluded that "the consequences for the international nuclear industry will be devastating," and "the international nuclear industry has been unable to stop the slow decline of nuclear energy" (Schneider 2012, 4). A report issued by the large banking firm UBS predicted that at least 30 nuclear power plants were likely to be closed permanently as a result of public reactions to the Fukushima accident, all of them in Belgium, France, Germany, and Switzerland. (Indeed, on May 31, 2011, Germany announced that it would close all 17 of its operating nuclear reactors by the year 2022.) The UBS report concluded that "the Fukushima accident was the most serious ever for the credibility of nuclear power . . . casting doubt on whether even an advanced economy can master nuclear safety" (Lekander and Oldfield 2012).

Other Renewable Energy Sources

To many people, renewable energy sources, such as solar, wind, and water, seem like the great promise for the future of energy production throughout the world. But humans have been using some forms of renewable energy since time immemorial. For example, some of the earliest forms of water transportation depended on the ability of sails to capture wind energy to power boats. The first water wheels probably date to at least the second century BCE when they were used by the Romans to crush grain, tan leather, smelt iron, saw wood, and perform a number of other operations. Still, it was not until the 19th century that the modern age of renewable energy can be said to have gotten underway. In 1860, for example, French inventor Augustin Mouchot (concerned about possible future shortages of fossil fuels!) invented a device for capturing sunlight and using it to convert water to steam for use in running industrial devices. Only 16 years later, William Grylls Adams, professor of natural philosophy at London's King College, and his student, Richard Evans Day, invented a second solar device made of selenium that could convert sunlight directly to electricity. Today's solar devices owe their existence to one of these two forms of solar energy, *thermal solar,* in which the energy of sunlight is converted to heat, and *photovoltaic,* in which that energy is converted to electricity.

Interest in other renewable sources of energy blossomed also. The first commercial operation for using the energy of running water to generate electricity, for example, dates to 1880, when the Grand Rapids (Michigan) Electric Light and Power Company used the energy of the Grand River to operate a water turbine that produced electricity to light the Wolverine Chair Factory. Shortly thereafter, the first commercial windmill built to produce electric power was constructed by Charles F. Brush in Cleveland, Ohio, in 1888. And soon after that, the first commercial geothermal system was opened in Boise, Idaho, where steam from nearby hot springs was piped into town and used to heat 200 homes and 40 businesses. (That system remains in use today and provides heat to 55 businesses in the downtown area.) One other form of renewable energy, biomass, has perhaps the longest and most extensive history of any renewable resources. For many centuries, humans used biomass in the form of wood as their primary source of energy. Interest in using biomass largely disappeared during the 19th century, however, as coal, oil, and gas became the fuels of choice for residential, industrial, transportation, and other purposes. By the last few decades of the 20th century, interest in biomass energy grew once more. Researchers began to explore new and more efficient ways of burning woody waste, garbage, and other forms of biomass to produce heat directly, as well as converting biomass to a gaseous form (gasification or anaerobic digestion) that could then be burned. The conversion of certain forms of biomass, such as corn, to liquid fuels (ethanol) also became an increasingly popular technology in many parts of the world.

Yet another form of renewable energy about which little is probably known among the general public is energy from the oceans. That energy is available in two ways: by the collection of heat energy stored in the oceans when seawater captures sunlight, and by the conversion of mechanical energy in the oceans (in the form of tides and waves, for example) into electricity. Both types of technology have histories as old as those of other forms of renewable energy. The earliest patent for a machine for capturing the energy stored in waves was probably one issued to French engineer Pierre Gerard and his son in 1799. A number of inventors followed up on this beginning, with 340 patents for such devices having been granted in the United Kingdom alone between 1855 and 1973. Devices for converting the energy of tides into mechanical energy on land have even longer histories, dating at least to the sixth century CE at many locations along the Irish coast. The first modern tidal power plant opened in 1966 on the Rance River, in Brittany, France. The plant continues to operate today, producing 250 MW of power.

Renewable energy sources have a number of advantages, the most obvious of which is probably the fact that they are renewable. That is, unlike the fossil fuels, which, once used up, are probably gone forever, they are essentially available forever. Solar power will be available as long as the Sun shines, as will wind power and biomass, which depend on sunlight for their availability. Hydroelectric power and ocean power will also be available as long as it rains and there is moving water on the planet. Another significant advantage of renewable resources is that they tend to have modest environmental effects, or none at all. At a time when concerns about the environmental and health effects of the combustion of coal, oil, and natural gas are very high, such an advantage is of some importance.

Proponents of renewable energy sources often point to their potential economic benefits also. Instead of being located in a certain number of centralized places, such as coal mines and oil and gas wells and refineries, many forms of renewable energy can be installed almost anywhere that weather conditions are suitable. This means that small communities that would otherwise have no connection whatsoever with the production of coal, oil, and gas energy can become involved in the energy industry, with the financial and employment benefits that change could generate.

Renewable energy also has its disadvantages. For one thing, most forms of renewable energy require somewhat specific environmental conditions. Wind farms, for example, cannot be built in areas where there is relatively little wind on an ongoing basis. Hydroelectric plants can be built only along rivers and geothermal plants can be built only in regions where heat from the Earth's interior exists in quantities great enough and close enough to the surface to be captured. Solar energy is theoretically available anywhere in the world where the Sun shines (which is just about everywhere), although using that solar energy as a reliable source of heat or electricity is likely to be successful only where the Sun shines for significant parts of the year (i.e., not in northern Russia).

One of the strongest arguments against renewable energy has long been its high cost, especially compared to the cost of fossil fuels. Between 1947 and 1973, the price of crude oil hovered around $20 per barrel, and returned to that level again between 1987 and 1998. In the interim period, between 1973 and 1987, the price of crude oil soared, to as high as $75 per barrel, as it did once more in the early 2000s, when it reached more than $100 a barrel, at least briefly. At times when oil was inexpensive, renewable energy sources could not begin to compete with oil and other fossil fuels. Part of the reason was that most renewable resources were still in the stage of development, a period during which costs are often much higher than after they have been fully developed.

In any case, the long-term prospects of at least some types of renewable energy sources seem good. As improvements are made in technology, costs tend to drop. And as the cost of traditional fuels increases, a point is likely to be reached at which cost is no longer a disadvantage for solar, wind, geothermal, and other alternative technologies.

That point was reached in 2010 in the competition between solar and nuclear power. As recently as 2000, the cost of solar photovoltaic power was about six times as great as the cost of nuclear power, about 30 cents per kilowatt hour for solar compared to about five cents per kilowatt hour for nuclear. In the following decade, the cost of solar power continued to decline, whereas the cost of nuclear power began to increase. Finally, in 2010, the two trend lines crossed at about 17 cents per kilowatt hour, with projections suggesting that the cost of solar power by 2020 would be less than five cents per kilowatt hour, and that for nuclear power, closer to 35 cents per kilowatt hour.

In 2005, the Energy Analysis Office of the U.S. National Renewable Energy Laboratory released its projections as to future trends in the cost of a variety of alternative forms of energy. That analysis showed a sharp decrease in the installed cost of all forms of renewable energy, including solar, wind, geothermal, and bio-based ethanol. It predicted that the reduction in costs would range from about 50 cents per kilowatt hour in 1980 for wind to less than five cents per kilowatt hour in 2025; from 15 cents per kilowatt hour for geothermal sources in 1980 to about three cents per kilowatt hour in 2025; and from between $2 and $7 per gasoline gallon equivalent for bio-based ethanol to between $1.5 and $2 per gasoline gallon equivalent in 2025 ("Renewable Energy Cost Trends" 2012). One of the first pieces of data that seemed to confirm this trend was announced in November 2011, when the latest reports from Bloomberg Business and Financial News confirmed that, for the first time in history, corporations had committed more to the development of renewable energy resources—$187 billion in 2010—than to the development of fossil fuel resources—$157 billion.

This news was especially interesting in a time of severe economic downturn worldwide and failed efforts at reviving the Kyoto Protocol on the reduction of greenhouse gases (both of which might have been expected to mitigate against the actual trends). Whether these trends will continue into the future depends on a number of factors, not the least of which will be the cost of fossil fuels themselves at that point in history (Morales 2012).

References

"About Syncrude." http://www.syncrude.ca/users/folder.asp?FolderID=5617. Accessed February 16, 2012.

"Alaska Energy for American Jobs Act." U.S. House of Representatives. February 9, 2012. http://www.gpo.gov/fdsys/pkg/CRPT-112hrpt393/pdf/CRPT-112hrpt393.pdf. Accessed February 16, 2012.

Aleklett, Kjell. *Reserve Driven Forecasts for Oil, Gas, and Coal and Limits of Carbon Dioxide Emissions.* Joint Transport Research Centre, 2007. http://www.internationaltransportforum.org/jtrc/discussionpapers/DiscussionPaper18.pdf. Accessed February 16, 2012.

"Annual U.S. Natural Gas Marketed Production." U.S. Energy Information Administration. http://www.eia.gov/dnav/ng/hist/n9050us2a.htm. Accessed February 16, 2012.

Canada's Energy Future: Energy Supply and Demand Projections to 2035. National Energy Board. November 2011. http://www.neb-one.gc.ca/clf-nsi/rnrgynfmtn/nrgyrprt/nrgyftr/2011/nrgsppldmndprjctn2035-eng.pdf. Accessed February 16, 2012.

Canadian Association of Petroleum Producers. *Statistical Handbook for Canada's Upstream Petroleum Industry.* Calgary: Canadian Association of Petroleum Producers, September 2011.

Colborn, Theo, et al. "Natural Gas Operations from a Public Health Perspective." http://www.endocrinedisruption.com/files/NaturalGasManuscriptPDF09_13_10.pdf. Accessed February 17, 2012.

"The Comprehensive Shell Remediation Plan for the Niger Delta." Offshore Energy Today.com. http://www.offshoreenergytoday.com/the-comprehensive-shell-remediation-plan-for-the-niger-delta/. Accessed February 16, 2012.

"Davos: Oil Discussion Round-upe [*sic*]." http://www.ifandp.com/article/001702.html. Accessed February 16, 2012.

Dickens, Charles. *Bleak House.* London: Bradbury and Evans, 1853.

Energy Watch Group. *Coal: Resources and Future Production.* http://www.energywatchgroup.org/fileadmin/global/pdf/EWG_Report_Coal_10-07-2007ms.pdf. Accessed February 16, 2012.

European Commission. "Air Quality Standards." http://ec.europa.eu/environment/air/quality/standards.htm. Accessed November 27, 2011.

"Fact Sheet: The Kyoto Protocol." United Nations Framework Convention on Climate Change. http://unfccc.int/files/press/backgrounders/application/pdf/fact_sheet_the_kyoto_protocol.pdf. Accessed February 17, 2012.

Fairley, Peter. "Alberta's Oil Sands Heat Up." *Technology Review* 114, no. 6 (2011): 52–59.

Hart, Nolan. "How Long until Peak Natural Gas?" http://doodlebugs.xomba.com/how_long_until_peak_natural_gas. Accessed February 16, 2012.

"Heavy Oil Resources of the Orinoco Oil Belt, Venezuela." Geology.com. http://geology.com/usgs/venezuela-heavy-oil/. Accessed February 16, 2012.

Helm, Dieter. "The Peak Oil Brigade Is Leading Us into Bad Policymaking on Energy." *The Guardian.* http://www.

guardian.co.uk/commentisfree/2011/oct/18/energy-price-volatility-policy-fossil-fuels. Accessed February 16, 2012.

Helms, Lynn. "Horizontal Drilling." https://www.dmr.nd.gov/ndgs/newsletter/NL0308/pdfs/Horizontal.pdf. Accessed February 16, 2012.

"Hydraulic Fracturing Overview." Keystone Energy Forum. http://www.keystoneenergyforum.com/article/hydraulic-fracturing-overview. Accessed February 17, 2012.

International Energy Agency. *World Energy Outlook 2010.* http://www.iea.org/Textbase/npsum/weo2010sum.pdf. Accessed February 16, 2012.

"Impact of Limitations on Access to Oil and Natural Gas Resources in the Federal Outer Continental Shelf." U.S. Energy Information Administration. http://www.eia.gov/oiaf/aeo/otheranalysis/aeo_2009analysispapers/aongr.html. Accessed November 22, 2011.

"Is Peak Oil Real? A List of Countries Past Peak." *The Oil Drum.* http://www.theoildrum.com/node/5576. Accessed November 21, 2011.

"Italy Needs to Face up to the Future." Power Engineering. http://www.powerengineeringint.com/articles/print/volume-16/issue-4/power-gen-europe-2008/italy-needs-to-face-up-to-the-future.html. Accessed February 16, 2012.

Klier, Thomas, and Joshua Linn. "Corporate Average Fuel Economy Standards and the Market for New Vehicles." http://www.chicagofed.org/digital_assets/publications/working_papers/2011/wp2011_01.pdf. Accessed February 17, 2012.

Laherrère, J. H. "The Hubbert Curve: Its Strengths and Weaknesses." http://dieoff.org/page191.htm. Accessed February 16, 2012.

Lekander, Per, and Stephen Oldfield. "Can Nuclear Power Survive Fukushima?" http://www.mp.se/files/242400-242499/file_242471.pdf. Accessed February 17, 2012.

Levine, Mark D. "Energy Efficiency in China: Glorious History, Uncertain Future." China Energy Group at LBNL. http://www.aps.org/units/maspg/meetings/upload/levine.pdf. Accessed February 17, 2012.

"Major Air Pollution Disasters." http://njcmr.njit.edu/distils/lab/Air_html/disaster.htm. Accessed February 17, 2012.

Montgomery, Carl T., and Michael B. Smith. "Hydraulic Fracturing: History of an Enduring Technology." *JPT Online* (2010). http://www.spe.org/jpt/print/archives/2010/12/10Hydraulic.pdf. Accessed February 17, 2012.

Morales, Alex. "Renewable Power Trumps Fossils for First Time as UN Talks Stall." Bloomberg. http://www.bloomberg.com/news/2011-11-25/fossil-fuels-beaten-by-renewables-for-first-time-as-climate-talks-founder.html. Accessed February 17, 2012.

"More Oil? Speculation Tantalizes Oil Explorers." Big Sky Business.com. http://archive.bigskybusiness.com/index.php?name=News&file=article&sid=1504. Accessed February 16, 2012.

Newby, Anna. "Arab Atomic Energy Agency Meets in Tunisia." Center for Strategic and International Studies. http://csis.org/blog/arab-atomic-energy-agency-meets-tunisia. Accessed May 30, 2012.

Newton, David E. *Chemistry of the Environment.* New York: Facts on File, 2007.

"Nuclear Power Reactors in the World." International Atomic Energy Agency. http://www-pub.iaea.org/mtcd/publications/pdf/rds2-26_web.pdf. Accessed February 17, 2012.

Osborn, Stephen G., Avner Vengoshb, Nathaniel R. Warner, and Robert B. Jackson. "Methane Contamination of Drinking Water Accompanying Gas-well Drilling and Hydraulic Fracturing." *PNAS* 108, no. 20 (2011): 8172–76.

"PDVSA Cuts and Delays Production Goals, Raises Capex Forecast." Setty's notebook. http://settysoutham.wordpress.

com/2010/09/16/pdvsa-cuts-and-delays-production-goals-raises-capex-forecast/. Accessed February 16, 2012.

Rembrandt. "Is the United Kingdom Facing a Natural Gas Shortage?" *The Oil Drum* (2012a). http://www.theoildrum.com/node/6113. Accessed February 16, 2012.

Rembrandt. "Jevons' Coal Question: Why the UK Coal Peak Wasn't as Bad as Expected." *The Oil Drum* (2012b). http://www.theoildrum.com/node/8241. Accessed February 16, 2012.

"Renewable Energy Cost Trends." Energy Analysis Office. http://www.nrel.gov/analysis/docs/cost_curves_2005.ppt#1. Accessed February 17, 2012.

Schneider, Mycle. "Nuclear Power in a Post-Fukushima World: 25 Years After the Chernobyl Accident." http://www.worldwatch.org/system/files/NuclearStatusReport2011_prel.pdf. Accessed February 17, 2012

"SD-PAMS Data Base." World Resources Institute. http://projects.wri.org/sd-pams-database. Accessed February 17, 2012.

"Sulfur Treaty." TED Case Studies. http://www1.american.edu/ted/SULFUR.HTM. Accessed February 17, 2012.

"3 to 4.3 Billion Barrels of Technically Recoverable Oil Assessed in North Dakota and Montana's Bakken Formation—25 Times More Than 1995 Estimate." U.S. Geological Survey. http://www.usgs.gov/newsroom/article.asp?ID=1911. Accessed February 16, 2012.

Toro, Francesco. "The Red Apertura and the Forgotten Art of Disarming Atomic Bombs." *Caracas Chronicles.* http://caracaschronicles.com/2011/09/26/the-red-apertura-and-the-art-of-disarming-atomic-bombs/. Accessed February 16, 2012.

2009–2010 IPAA Oil and Gas Producing Industry in Your State®. Washington, DC: Independent Petroleum Association of America. http://www.scribd.com/

doc/78258474/2009-2010ipaaopi-Report. Accessed February 17, 2012.

2010 Survey of Energy Resources. World Energy Council. http://www.worldenergy.org/documents/ser_2010_report_1.pdf. Accessed February 16, 2012.

U.S. Energy Information Administration. "Analysis of Crude Oil Production in the Arctic National Wildlife Refuge." 2012a. http://205.254.135.7/oiaf/servicerpt/anwr/back ground.html. Accessed February 16, 2012.

U.S. Energy Information Administration. "Table 13. Natural Gas Supply, Disposition, and Prices." 2012b. http://www.eia.gov/oiaf/aeo/excel/aeotab_13.xls. Accessed February 16, 2012.

U.S. Environmental Protection Agency. "1970–2011 Average Annual Emissions, All Criteria Pollutants." http://www.epa.gov/ttn/chief/trends/trends06/nationaltier1upto2011 basedon2008v1_5.xls. Accessed February 16, 2012.

Vallero, Daniel A. *Paradigms Lost: Learning from Environmental Mistakes, Mishaps and Misdeeds.* London: Butterworth-Heinemann, 2005.

"Weekly Rig Count." Petrodata. http://www.ods–petrodata.com/odsp/weekly_rig_count.php. Accessed November 22, 2011.

"What Chemicals Are Used?" FracFocus. http://fracfocus.org/chemical–use/what–chemicals–are–used. Accessed November 25, 2011.

"Why We Need More Development on Government Lands and Offshore." American Petroleum Institute. http://www.api.org/policy-and-issues/policy-items/exploration/why_we_need_more_development.aspx. Accessed February 16, 2012.

Winston, Ken. "Tar Sands Pipeline Update." Sierra Club—Nebraska Chapter. http://sierranebraska.org/2012/02/13/tar-sands-pipeline-update-february-13-2012/. Accessed February 2012.

"World Annual Oil Production (1900–2009) and Peak Oil (2010 Scenario)." The Geography of Transport Systems. http://people.hofstra.edu/geotrans/eng/ch5en/appl5en/worldoilreservesevol.html. Accessed February 16, 2012.

World Coal Association. "Where Is Coal Found?" http://www.worldcoal.org/coal/where-is-coal-found/. Accessed February 16, 2012.

World Shale Gas Resources: An Initial Assessment of 14 Regions Outside the United States. Washington, D.C.: U.S. Energy Information Administration, April 2011.

Wulfinghoff, Donald R. "The Modern History of Energy Conservation: An Overview for Information Professionals." http://www.energybooks.com/resources/modern_history_of_energy.pdf. Accessed February 17, 2012.

Yergin, Daniel. "There Will Be Oil." http://online.wsj.com/article/SB10001424053111904060604576572552998674340.html. Accessed February 16, 2012.

BUCYRUS
495HF
797B
797B
118

3 Perspectives

Discussions about the world energy crisis evoke strong feelings across a broad range of opinions. Some individuals and organizations argue either for or against any number of propositions (such as, "Is 'peak oil' a valid concept?"), or for or against various technological solutions for the world's energy problems (such as "Should we or should we not pursue the world's tar sands assets?"). This chapter presents a number of position statements along these lines. The author's original goal was to obtain balanced statements both for and against a variety of positions. But space does not allow for a complete variety of essays, and a number of organizations and individuals invited to present such statements did not respond or declined to submit an essay. The essays included here, however, do present some thoughtful and cogent position statements on some of the most serious energy issues facing the world in the early 21st century. The author expresses his heartfelt appreciation to the contributors.

THE CASE FOR DEVELOPMENT OF OIL SHALE IN THE UNITED STATES

R. Glenn Vawter

The expression *energy crisis* has been used loosely over the years. But in the 1970s we indeed had such an event. Supplies

---

Mining trucks carry loads of oil-laden sand after being loaded by huge shovels at the Albian Sands oils sands project in Ft. McMurray, Alberta, Canada. (AP Photo/Jeff McIntosh)

of crude oil were cut off by the petroleum cartel Organization of Petroleum Exporting Countries (OPEC) and gasoline supplies became limited. Lines at gas stations were common for several weeks. The United States had become dependent on foreign supplies of oil—a situation that persists in 2012. Following the OPEC embargo of the 1970s, the U.S. Congress enacted legislation to encourage the domestic production of energy. Since our conventional oil fields were on the decline the government encouraged the development of unconventional fuels such as oil shale, coal, and tar sands since these energy resources existed in huge amounts in the United States. However, the crisis passed, oil prices dropped, supplies stabilized, and the initiatives to produce unconventional domestic sources of fossil energy were cancelled.

Although we have not seen a recurrence of the energy crisis of the 1970s, the potential for it to occur again looms over us. Since the 1970s each U.S. federal administration and Congress has taken different and inconsistent approaches to dealing with the issue. The nation continues to import over half of its petroleum, the demand for petroleum in developing countries is increasing, and world suppliers may struggle to keep up with the rising demand. Thus, nations will compete for supplies and the price we pay at the pump will continue to rise.

In light of this state of affairs, the United States needs to increase domestic supplies of energy from all economic and sustainable sources whether they are fossil or renewable. Our lifestyle depends on reasonably priced and secure supplies of petroleum.

Many in this country have been led to believe that renewable sources of energy can replace fossil sources over a short time frame. That is not the case. Most of the petroleum consumed in this country is refined into transportation fuels used in our cars, trucks, trains, and planes. We will be dependent on fuel from petroleum for decades even with the best efforts to convert to other sources. Most renewable sources, such as wind and solar, produce electric power and not transportation fuels. Yes, electric cars can help, biofuels can make a contribution, and natural gas–powered vehicles are important in the overall mix. However, increasing domestic supplies of petroleum can have a much larger impact on reducing dependence on foreign energy supplies and the stabilizing prices.

Oil shale is one of the most promising sources of unconventional petroleum. The oil shale resource is huge. The U.S. Geological Survey has estimated that over four trillion barrels of oil is contained in oil shale resources in the states of Colorado, Utah, and Wyoming (Bartis 2011; U.S. Geological Survey 2011). How much is recoverable is debatable, but something in the range of 25 percent or 1 trillion barrels is not an unreasonable assumption. There are additional oil shale resources in eastern states.

Heat is required to release oil and gas from oil shale. This is accomplished either by mining and surface processing, known as *ex situ*, or by the use of oil field techniques, known as *in situ*. This differs from conventional oil and gas, which are drawn directly from oil reservoirs without heating.

Progress has been made in oil shale research and development in the United States and overseas in recent years. Commercial production of oil and gas from oil shale is ongoing in Estonia, China, and Brazil. Experimental plants were operated in the United States in the 1970–1980s before the collapse of oil prices and a loss of interest in unconventional fuels by the U.S. government. Technologies for *ex situ* and *in situ* processes are much more advanced than those anticipated for use and tested in the 1970–1980 era. The approach to project development is also different than the rush to commercialization of that era. The current development approach is deliberate, staged, and focused on finding technical solutions that are economically viable and environmentally and socially acceptable. Projects will meet strict regulatory standards in order to proceed.

Recent shale gas and tight oil development in Texas, North Dakota, Pennsylvania, and elsewhere has created a new industry that is taking hold around the world. The presence of gas and oil in shale rocks, not to be confused with oil shale, has been known for decades, but it is the development of new horizontal drilling and stimulation technologies that have made the resource economically attractive. Oil shale is on the brink of a similar renaissance as new technologies are developed and older technologies are optimized for the production of shale oil.

The facts about oil shale are often obscured by myths that have been perpetuated for decades by those who either wish to discredit the potential of this huge energy resource, or are simply misinformed. Oil shale has a high energy content compared to other fossil resources, and developers estimate that at least three units of energy can be produced for one unit of energy input. Water is available for the development of the resource (Bunger 2010; U.S. Department of Energy 2005). Developers realize the value of water in the arid west, are reducing water requirements to a minimum, and are seeking alternate sources. Lastly, carbon dioxide produced during the production of shale oil can be captured and sequestered. These are but three facts about oil shale that are often misconstrued.

There are significant challenges facing a private developer hoping to move toward building a commercial oil shale project in the United States. Technology risk is one challenge. *In situ* technologies have not yet been tested at a scale that provides adequate confidence to design a commercial project. *Ex situ* technologies commercialized abroad have advanced beyond the experimental stage, but still need to be demonstrated using U.S. oil shale. Companies are investing in their proprietary technologies without U.S. government funding to solve these technical problems.

Technology development is not the most significant challenge. The lack of a consistent U.S. federal policy for oil shale leasing and regulation, similar to what already exists for other minerals and oil and gas, is restraining long-term investment in the development of the resource. States that control oil shale resources also have differing policies toward its development. The federal government controls 80 percent of the western oil shale resource, so it has a unique and vested interest to help facilitate an oil shale industry if it can be done in an environmentally and a socially acceptable manner. Lastly, fossil energy development needs to be encouraged on a national basis to start to reduce the political and economic burden of continuing to import so much petroleum.

References

Bartis, James T. "The Roadmap for America's Energy Future." Testimony presented before the House Energy and Commerce Committee, Subcommittee on Energy and Power, 1. June 3, 2011. http://www.rand.org/content/dam/rand/pubs/testimonies/2011/RAND_CT363.pdf. Accessed February 14, 2012.

Bunger, James W. D. Letter to The Honorable Mike McKee, Commissioner, Unitah County, Utah, Energy Efficiency and Water Demand sections. September 2010. http://www.energy.utah.gov.government/strategic_plan/docs/publiccomments/jameswbunger10142010.pdf. Accessed February 14, 2012.

U.S. Department of Energy, Office of Petroleum Reserves. "Fact Sheet: Energy Efficiency of Strategic Unconventional Resources." [2005.] http://fossil.energy.gov/programs/reserves/npr/Energy_Efficiency_Fact_Sheet.pdf. Accessed February 14, 2012.

U.S. Geological Survey. "Fact Sheet: Assessment of In-Place Oil Shale Resources of the Green River Formation, Greater, Green River Basin in Wyoming, Colorado, and Utah." Fact Sheet 2011-3063, Resource Summary Section. June 2011. http://pubs.usgs.gov/fs/2011/3063/pdf/FS11-3063.pdf. Accessed February 14, 2012.

R. Glenn Vawter is executive director of the National Oil Shale Association, a nonprofit group dedicated to providing factual information about oil shale. His background includes engineering, management, and executive positions in the energy sector. He graduated from the Colorado School of Mines and attended the Harvard Business School's Advanced Management Program.

OFFSHORE OIL-DRILLING PRIMER FOR CONCERNED PEOPLE OF ALL AGES

Jan Lundberg

Petroleum is a natural substance beneath the surface of the Earth that is found in two forms: crude oil and natural gas. Oil has been of inestimable importance in the transformation of both rural and urban living and of the landscape. There is more paved surface, mostly of asphalt made from oil refining, in the United States than in areas designated as official "wilderness areas."

The amazing applications of petroleum and its popularity, as oil discoveries and extraction approached their all-time peak, have started to be questioned. So far this questioning is often in the form of attempting different propulsion technologies for cars, even though they would still use the same road system that contributed to unsightly, inefficient urban sprawl. There are two main reasons for this shift in attitude about oil: (1) environmental damage, such as from the infamous BP *Deepwater Horizon* blowout in the Gulf of Mexico in 2010 and (2) the beginning of the end of the Age of Oil, since that maximum supply has peaked globally (many say in 2005). So, society needs to decide what is truly sustainable, and not just green from a corporate marketing standpoint.

First, let us consider the ecological crisis to which oil has contributed so much. We can trace the beginning of new awareness about oil's danger to 1969, when Santa Barbara, California, beaches suddenly became fouled with toxic oil from an offshore oil rig's blowout. This made international news, setting into motion not just strong feelings by passive animal lovers seeing a tragic die-off of sea birds, but also a historic movement to protect our whole natural environment. All that was needed was a catalyst, and the Union Oil Company's accident off the coast of Santa Barbara was it. The first Earth Day followed months later in 1970.

Unfortunately, the problem of oil spills—whether from ships going aground or oil rig failures—has never been solved. The approach of government and its influential friends in the oil and shipping business is only to mitigate or address pollution, not stop it. So, one reform has been to double-hull the oil tankers so that wrecks do not so easily result in disasters for the ecosystem—and bad publicity for the oil industry.

Stopping offshore oil drilling, at least in one's "viewshed" (what people look out on and see), is a common sentiment, but much less common is the opposing of oil exploration all together. Even less common is a lifestyle of deliberately boycotting oil. Whether people will decide to end oil dependence voluntarily or wait until finding themselves without oil or refined products such as gasoline, is a question seldom discussed.

When young people consider offshore drilling of oil as something either necessary or bad, it is often without much consciousness of the driving lifestyle, of the plastic plague, oil wars, or political corruption of the government by industry interests. Some oily practices right under our nose, such as driving to the store for just one item—when one could have walked, skateboarded, or bicycled—have their days numbered. Even non-oil-propelled cars are included in this assessment, for reasons of scalability of alternative fuels, the poorer net energy from them compared to the much depleted cheap oil of yesterday, and the still oil-dependent aspects of the whole car infrastructure of today's industrial society.

When someone who imagines herself or himself as an environmentalist and objects to offshore oil drilling while clinging to a lifestyle centered around the car, there is clearly a need for some deeper thinking. Educating ourselves on the issues of energy is crucial, but we need to pay attention to our feelings as well: violence and pollution caused by cars are essential to recognize if we are a society making conscious decisions. We need to ask ourselves questions such as, "Can nature withstand more and more oil spills?" And, "Is there another way to live that is less costly, more efficient, and kinder to Mother Earth?"

Offshore oil drilling is getting more intensely pursued because the easy oil extraction from shallow wells on land is dwindling fast. In fact, to try to drill for oil in difficult conditions—say, roughly one mile under the surface of the sea, as the BP Macondo well was, or in the harsh Arctic—means the best oil reserves are depleting. And the new, expensive wells do not yield as much net energy for the energy applied, compared to wells' former average efficiency. This ratio is worsened dramatically with the potential for massive spills, as with the BP Macondo well that released nearly 5 million barrels of oil and about 100 million standard cubic feet per day of natural gas (Congressional Research Service).

Despite the historic oil spills of Santa Barbara Channel and BP Macondo, and other spills or blowouts around the world, the U.S. government is all for more drilling. One reason might be that BP has been the biggest U.S. military contractor for oil to Afghanistan. President Barack Obama has not advocated cutting our oil use except to call for eventual renewable energy dependence, even though there are many ways to halt wasteful oil consumption now.

Depending on how an offshore oil rig is defined (there are various types, and few rigs are deepwater), and depending on the fluctuation of utilization of rigs and platforms, there are a few dozen to several dozen offshore oil drilling rigs and platforms near the United States. There are another few hundred elsewhere in the world. There are few inspectors and little oversight, and not all accidents or catastrophic events are reported.

It is for the new generation of aware world citizens to decide if this state of affairs will be tolerated or opposed and terminated. Critical thinking demands that we see that the world environment is a closed system. Therefore, a spill, blowout, or fire on land in one part of the world ultimately affects the health of the oceans far away, in part because of the rapid acidification of sea water due to carbon-dioxide emissions. And industry activity at sea affects inland ecosystems and climate. Offshore oil extraction is just one form of fossil fuels industrial activity degrading the biosphere. It profits few people, provides conveniences to more, but it affects everyone and everything in the long term.

Here are some sources of information on pollution from offshore oil pollution and what to do about it:

- Committee Against Oil Exploration: http://www.culturechange.org/cms/content/view/637/66/
- World Oil Reduction for the Gulf: http://worldoilreduction.org/
- Journalism exposing offshore oil's effects: http://motherjones.com/blue-marble/2012/01/report-white-house-pressured-scientists-underestimate-bp-spill-size
- Congressional Research Service: http://www.crs.gov/pages/Reports.aspx?PRODCODE=RL33705&Source=search.
- Report: White House Pressured Scientists to Underestimate BP Spill Size: http://motherjones.com/
- A recent half-hour documentary on Santa Barbara's remembrances of 1969, Janet Bridgers' film *Stories of the Spill* (trailer): http://www.youtube.com/watch?v=SDXqEk9W-Cw

Jan Lundberg learned about the oil industry when working for Lundberg Survey, a firm that predicted the Second Oil Shock in 1979. He left oil industry work in 1988 to get back to his college activism for a cleaner environment, and now works with the Sail Transport Network.

PROSPECTS FOR OIL SANDS IN THE 21ST CENTURY

Kathryn Marshall

The 21st century dawned on a world heavily dependent on fossil fuels. It may very well close on a world entirely free of them.

In between, there must be a massive shift in how we supply and use energy. Canada's oil sands will play an important part in that transformation.

It is one of the ironies of our desire to move to a petroleum-free future that it will take a great deal of petroleum to get us there. Statistics from the International Energy Agency tell us that for all the subsidies governments around the world have pushed onto alternative, renewable energy, and the immense psychological, physical, and financial resources that have been focused on finding ways to make wind and solar power economically competitive with fossil fuels, as of 2006, solar and wind power still comprised less than 1 percent of the world's energy supply mix; oil, gas, and coal provide 80 percent of the world's power.

Imagining new ways to reshape and improve our world is important. Just as important is thinking realistically about how we can ensure it comes to pass. When we think realistically about how much energy the world utilizes every day, and how the overwhelming majority of it comes from fossil fuels, we quickly understand that this is not something that will change quickly. It will take decades to find ways to effectively and economically replace carbon-emitting energy with carbon-free energy and many decades beyond that to rearrange our industries, our transportation networks, our power grids, and our lives, around new energy supplies.

If there is a carbon-free future ahead, we must rely on fossil fuels to take us there, to bridge the gap between our traditional energy infrastructure and the new, carbon-free society.

As modern, powerful economies blossom in China and India, the demand for energy is only growing. It will not stop, let alone diminish. At the same time, the availability of easily accessible and plentiful oil and gas is decreasing. In response, the energy industry has created innovative ways to tap previously uneconomical, unconventional deposits of oil and gas. Today, oil and gas trapped in shale and sand that was unavailable before, or too expensive to exploit, are suddenly opening up entirely new and vast energy stores.

They have not come without controversy: environmental nongovernmental organizations, fixated on the vision of a carbon-free future, worry that freeing up immense new stores of fossil fuels will only enhance our dependency on carbon-emitting energy. But our dependency is already a given. It's here. It's a fact. Today. In reality, we cannot stop using fossil fuels overnight. We must rely on them, because they are, currently, all we have to rely on. Until researchers and inventors develop a means to affordably run our economies on something other than fossil fuels, no government anywhere would welcome the gradual drying up of fossil fuel sources, risking shortages, economic havoc, and ultimately the impoverishment of citizens whose quality of life depends so heavily on heat, transportation, refrigeration, medicines, food, and so many other essentials and extras that are made possible by a predictable and a robust supply of affordable energy.

Governments, particularly in democratic, liberal societies like the United States, will have a choice to make in which sources of oil and gas they rely on to help them bridge the gap from a carbon-heavy society to a carbon-free one. And, increasingly, consumers and the voting public in democratic, liberal societies are seeking to make ethical choices about the products they use and the way they use them. The Fair Trade Coffee phenomenon is a good example: major coffee retailers from Starbucks to Wal-Mart have agreed to source coffee beans from a program that guarantees a minimum, fair income to coffee growers. The pressure on companies such as Nike to stop using child labor to produce consumer goods is another good example of how people are giving more thought, and more weight, to the ethical dimension of what they consume.

For decades, the world has been dependent largely on a handful of major oil exporters for their energy supply. Although these countries have used their oil power to create a price-fixing cartel, known as the Organization of Petroleum Exporting Countries (OPEC), even manipulating supply to deliberately trigger economically ruinous oil shocks, as OPEC did in 1973 to influence the U.S. Middle Eastern policy, and although the members of OPEC are ranked as some of the worst countries on the Earth for women's rights, democracy, human rights, and protection for minorities, they have kept countries such as the United States and European countries as their customers. The reason is simple: there were simply no better oil supply alternatives for the United States and Europe. Regimes that imprison democrats, stifle free speech, execute homosexuals, degrade women, and fund terrorism and war, have been able to keep their police states comfortably funded with the money they have collected from energy users in liberal, democratic, and peace-loving countries.

The innovations that have unleashed huge new deposits of oil sands oil in Canada means Americans will finally have a choice in supporting the ethics of where they buy their oil from, even as they're moving gradually towards an oil-free future. The size of the reserve of oil inside Canada's oil sands is the third largest in the entire world. The Canadian Association of Petroleum Producers reports that the oil sands contain 170 billion barrels of recoverable oil. The United States imports roughly 12 million barrels of oil a day, according to the U.S. Energy Information Administration. At that rate, in theory at least, Americans could rely solely on Canadian oil sands oil for the next 40 years, and not have to buy another drop from the dictators in Saudi Arabia or the hostile Hugo Chavez regime in Venezuela.

Will the United States be free of its oil dependency in 40 years? Probably not—at least not entirely. But it will be much closer to achieving a future free of fossil fuels. Its consumption of fossil fuels will be lower. And as that inevitable, carbon-free day approaches, Americans are now faced with the choice of which oil sources they will use in the meantime, to bridge the gap between traditional energy arrangements and a renewable energy society. Canada's oil sands have, for the first time, offered Americans a reliable supply of oil from a neighbor and ally who values human rights, democracy, gender equality, and protection for minorities just as much as Americans do. For the first time in history, Americans will have a plentiful, stable, and nearby source of oil that does not require them to sacrifice their ethics to consume it. As the United States—and the world—spend the next few decades finding the way to an economy powered strictly by renewable energy, Canada's oil sands will provide the ethical source of energy that will get them there.

Kathryn Marshall is a columnist, blogger, commentator, and spokesperson for EthicalOil.org. Kathryn has worked on political campaigns across Canada and has worked for the Fraser Institute, Canada's largest think tank. Kathryn holds a degree in women's studies and is currently completing a law degree.

EthicalOil.org is a grassroots advocacy organization that encourages people, businesses, and governments to choose ethical oil from Canada, its oil sands, and other liberal democracies. Unlike conflict oil from some of the world's most politically oppressive and environmentally reckless regimes, ethical oil is the fair trade choice in oil.

DRILLING IN THE MARCELLUS SHALE

Michael Pastorkovich

The Marcellus Shale is a geological formation consisting of mostly black shale and is believed to hold as little as 84 trillion cubic feet (TCF) of natural gas, according to recent U.S. Department of Energy estimates, or as much as 484 TCF according to Pennsylvania State University geoscience professor Terry Engelder (Orcutt 2011). This structure, which stretches from southern New York State, through three-quarters of mostly northern and western Pennsylvania, to eastern Ohio, most of West Virginia, and into the far western end of Virginia, formed nearly 400 million years ago and rested peacefully with its treasure trove of gas beneath the surface of the Earth until the early 2000s when the shale, so to speak, hit the fan.

Drilling for shale gas was relatively uncommon until the early years of our present century because the drilling techniques available until then made extraction of the gas a difficult and a costly proposition. The process for extracting gas from shale is called hydraulic fracturing or, more commonly, fracking, and it involves boring into the shale, sinking a shaft, and then injecting massive amounts of water and sand along with a brew of various chemicals, some of them highly toxic, under high pressure in order to create fractures and fissures to release the gas, which is then pumped to the surface. The traditional method of extracting natural gas involves drilling a vertical shaft, but because the fractures created by fracking do not spread very far from the point of impact, vertical drilling necessitates blanketing a huge area with wells in order to exploit the gas potential of a given region, which, in turn, entails significant construction costs. However, there were major improvements at the turn of this century in the technique called horizontal drilling, which allows for tilting the drill, once it has entered the shale, at an angle of 90 degrees so that drilling can continue sideways for up to a mile, thus accessing more gas with no additional construction. These improvements rendered shale drilling more profitable and, hence, more commercially viable in the eyes of gas companies. Thus has commenced an ongoing drilling boom in the Marcellus Shale. But in its wake, this boom has given rise to concerns on the part of an increasingly aware and worried public about the possibility of grave environmental damage as a result of the fracking process, and this worry has fostered, in some quarters, opposition to all drilling in the Marcellus. Politically, the issue of drilling has become one of the hottest in the states straddling the Marcellus.

The two strongest selling points put forward by the extraction companies and their political allies are that drilling in the Marcellus Shale will create thousands of jobs while at the same time advancing the United States along the road to energy independence. The fact that the Shale underpins the region of our nation often referred to as the rust belt, that is, an area that was once an industrial powerhouse but the manufacturing infrastructure of which has been long since dismantled and outsourced to cheap labor havens throughout the world, lends the mantra "jobs, jobs, jobs" a great deal of appeal. On the other hand, the idea that the United States must depend for its energy needs on nations that many Americans perceive as "hating us" and "hating our way of life" sometimes drives such people to the point of being willing to do almost anything to bring such a state of affairs to an end. Yet it is questionable whether the drilling will achieve either of these goals.

While it is indeed true that drilling in the Marcellus will create jobs, many of these jobs, such as well construction and soliciting property owners to sign drilling leases, are temporary. On the other hand, the gas companies often bring professionals from places like Texas and Oklahoma to the east in order to perform the actual work of drilling, fracking, and extracting, rather than training locals to perform those tasks. Furthermore, the wells themselves have a finite lifetime, although just how long or short is a matter of dispute. The extraction industry likes to proffer figures like 30–40 years, whereas researchers like Arthur Berman of the Association for the Study of Peak Oil & Gas–USA finds "the average commercial life for horizontal wells . . . about 7.5 years" (Berman 2009), a considerably more modest estimation. In any event, there is no doubt that sooner or later the gas will be tapped out, and when that happens, the workers will leave and the business that serviced them will disappear as well. The boom will invariably be followed by a bust.

As for energy independence, the term is normally used to designate ending our dependence on foreign sources for petroleum, and while natural gas can substitute for oil as a source of heating or for electrical power generation, it cannot replace oil in three of its most widespread and significant uses: (1) in the manufacture of gasoline, diesel fuel, and jet fuel; (2) in the manufacture of engine and machine lubricants; and (3) in the manufacture of plastics. This being the case, the role natural gas will play in the quest for energy independence is bound to be limited.

There is a third advantage that pro-drilling advocates sometimes like to tout, especially with the environmentally aware segment of the public, and it is that natural gas has a much lower carbon footprint than either oil or coal and makes an ideal transition fuel from fossil to green sources of energy. There is merit to this claim, but only if we compare the amount of greenhouse gases released in the burning of these three fuels. By that standard, natural gas releases nearly 30 percent less carbon dioxide than oil and nearly 50 percent less than coal. However, there is a big difference between conventional gas drilling and the fracking of shale to obtain natural gas. There is evidence that fracking releases 40–60 percent more methane into the atmosphere than conventional gas drilling, and methane traps 20–25 percent more heat in the atmosphere than carbon dioxide (Fischetti 2012). So, at best, it is highly questionable how beneficial the use of shale gas would be in the fight against global warming.

On the other hand, public anxiety about the environmental consequences of shale gas drilling and, specifically, the fracking process encompasses primarily three areas: water contamination, air pollution, and noise pollution. Of these, the potential for water contamination is the greatest concern, since such contamination can occur in a variety of ways.

The Marcellus Shale lies anywhere from nearly a mile to over a mile beneath the surface of the earth and, because of this, in some areas drills and piping have to pass through aquifers or other groundwater in order to reach the shale. As the fracking fluid passes through the pipes, contamination can occur from leakage. Although the two principle ingredients of fracking fluid are water and sand, from 1 to 2 percent consist of over 40 different chemicals added to perform such functions as preventing the growth of aquatic organisms around the pipes and the drills, to reduce corrosion and to lower friction. Some of these chemicals, such as benzene, toluene, and formaldehyde, are known carcinogens. Now, 1–2 percent might not sound like a lot, but when you consider that 3–8 million gallons of water are used for each fracking procedure, 1–2 percent can mean anywhere from 30,000 to 160,000 gallons of some pretty toxic stuff.

Furthermore, while some of the solution remains underground, some returns to the surface when the gas is pumped up, and this wastewater contains not only the chemicals added for fracking but other substances, including radioactive substances, from dissolved rock. The Pennsylvania Department of Environmental Protection has estimated that the state's oil and gas wells would produce 19 million gallons of this waste per day by the year 2011 (Sustainable Otsego 2009). This wastewater must be disposed of, one way or another. One way is to store it in holding ponds, which can develop leaks or can overflow in heavy rains, polluting soil and nearby sources of fresh water. Another way, called the "deep injection well disposal", involves pumping millions of gallons of toxic waste into subsurface reservoirs, such as abandoned gas wells, but this, again, runs the risk of contaminating groundwater. Also, this deep injection method has been shown to cause earthquakes in places like Ohio, which are not normally thought of as earthquake-prone. So far, the quakes have been minor. Finally, drilling companies often recycle the waste, which means hauling the millions of gallons of this toxic brew to special treatment facilities (regular municipal water treatment plants are not capable of doing the job) to remove various pollutants and then return this treated water back into the rivers. There is evidence, however, that this recycled water contains high levels of bromide which, when exposed to the chlorides commonly used to disinfect drinking water, can form compounds known as trihalomethanes, which have been linked to birth defects and cancer (Hopey 2011).

One other water contamination issue of note is the fairly common escape of methane gas (natural gas is mostly methane) into groundwater supplies that mixes with well water in rural areas where a great deal of the drilling is done. This was likely the cause of the flaming faucet phenomenon that made such an impression in the documentary film *Gasland*, in which a Colorado resident of a shale gas drilling area turns on the tap and, using a cigarette lighter, produces a fireball. While methane in drinking water is not poisonous in and of itself, there is, of course, the danger of explosion and fire from the leaking gas.

In addition to the release of methane into the atmosphere mentioned previously, many drilling sites are located in places far from ready access to the millions of gallons of water needed for fracking. Consequently, all of this water has to be trucked in (and wastewater trucked out). These trucks emit tons of diesel fumes and particulates into the atmosphere.

As for noise pollution, one man living in the vicinity of a large gas compressor station describes his experience as "[n]oise that is constant . . . [n]oise that sounds like a jet plane circling overhead for 24 hours a day . . . [n]oise that drives people to the breaking point." Of course, this is not a problem for people who do not live near compressing stations, but for those who do it means that their homes, which may be where the bulk of their money has been invested, have become unsalable nightmares (Sustainable Otsego 2009).

Such concerns and problems, some of them potentially catastrophic, have led many people, both activists and just plain folks, to call for a total ban on fracking and shale drilling in the Marcellus formation. Others call for at least a moratorium, that is, a ban on drilling until objective, scientific studies can establish that the process can be done safely and until appropriate laws are enacted to ensure that safe practices are diligently followed, that no corners are cut, and that, if they are cut, the guilty parties are harshly penalized.

The moratorium route seems to me to be a most sensible approach. It is the rush to drill that strikes me as being irrational. The gas has been in the ground for millions of years. It will still be there when the scientific studies have been completed, regardless how long they take. But the key point is that humankind can survive and for thousand of years in the past has survived without the use of natural gas and other fossil fuels. But humankind cannot survive without fresh, drinkable water.

References

Berman, Arthur. "Commentary and Reports." First Enercast Financial. http://www.firstenercastfinancial.com/commentary/?cont=3193. Accessed May 20, 2012.

Fischetti, Mark. "Fracking Would Emit Large Quantities of Greenhouse Gases." *Scientific American,* January 20, 2012.

Hopey, Don. "Bromide: A Concern in Drilling Wastewater." *Pittsburgh Post-Gazette,* March 13, 2011.

Orcutt, Mike. "How Much U.S. Shale Gas Is There, Really?" *Technology Review,* August 31, 2011.

Sustainable Otsego. White Paper prepared for Congressman Michael Arcuri, November, 2009.

Michael Pastorkovich has a degree in philosophy and has spent the majority of his working life in the telecommunications industry. He is a long-time member of the Sierra Club and is active in the environmental movement along with other social causes.

THE PROBLEM OF CANADA'S TAR SANDS

Aaron Sanger

Today the United States imports more petroleum products from Canada than from any other foreign country. And most of those petroleum products from Canada are made from raw material extracted in the northern part of the Canadian province of Alberta—from a place called the Tar Sands. The geographic extent of Canada's Tar Sands is very large, covering areas in northern Alberta and also in the neighboring province of Saskatchewan that together equal the size of the U.S. state of Florida.

According to the oil industry, the potential for petroleum products from Canada's Tar Sands is second only to estimated conventional oil reserves in Saudi Arabia. Drawn by the tremendous profits that can be made over the next several decades as peak oil pushes the global price of oil higher, the oil industry has invested hundreds of billions of dollars in the Tar Sands.

How Is Oil Connected to the Tar Sands?

Many in Canada talk about the Tar Sands as *oil sands*, but what they are talking about is not yet oil when it is extracted from the ground as a tarry substance. The technical name for this substance is bitumen. Chemically, bitumen is a highly sulfurous hydrocarbon that lacks sufficient hydrogen to function as petroleum. Before it can be used to make many types of petroleum products, such as gasoline or diesel, a refinery must first turn bitumen into a type of synthetic oil.

How Are Tar Sands Extracted?

Most of the Tar Sands material coming out of Alberta today is extracted using a form of strip mining for which the first step is removal of boreal forest. The next step is removal of, on average, four tons of earth for every barrel of bitumen. Huge trucks move this earth-containing bitumen to a processing facility where the earth and bitumen are separated by hot water. The water is heated using natural gas, a cleaner form of hydrocarbon energy than petroleum.

Only 20 percent of the bitumen in Canada's Tar Sands can be strip-mined. The other 80 percent must be extracted by drilling holes in the ground, injecting steam (again produced by heating water with natural gas) to liquefy the bitumen and suck it out of the ground. This method requires roads, pipelines, and other infrastructure that break up the boreal forest into small islands. The roads and pipelines, in turn, require draining the muskeg or wet, boggy terrain that is required for the kind of boreal forests that grow in the Tar Sands area.

What Are the Impacts on Environment and Communities?

Extracting Tar Sands, whether by strip mining or drilling, damages or destroys boreal forests. Based on existing leases for Tar Sands extraction, strip mining and drilling could damage or destroy an area of boreal forest the size of the U.S. state of Maine or the country of Scotland, an area of more than 32,000 square miles.

Wetlands can also be destroyed in ground operations required either for strip mining or drilling to extract Tar Sands. In the strip-mining phase alone, it is possible that 770 square miles of wetlands will be permanently eliminated.

Because of the severe potential impacts on forests, wetlands, and other habitat, Tar Sands operations endanger herds of woodland caribou that are already facing a high risk of extinction. An estimated 200 million boreal songbirds could be lost over the next 30–50 years because of Tar Sands extraction.

Tar Sands operations consume prodigious quantities of energy and water—as much as one-third of the barrel equivalent in energy and up to three barrels of water for every barrel of bitumen produced. The greater energy consumption required by Tar Sands extraction is one reason it is so polluting. For example, per barrel produced, Tar Sands oil generates three to five times more greenhouse gases than conventional oil.

The way water is used by Tar Sands strip mining is also a severe pollution problem. Although the industry reuses water as much as possible, ultimately it becomes so saturated with toxins that it cannot be reused again. This toxic waste is dumped into huge open pits that now cover an area the size of Washington, D.C. Today all of these pits contain high concentrations of lethally toxic chemicals. The drilling method does not require dumping toxic waste in open pits; but the trade-off is that highly toxic chemicals are unleashed in extremely hot water left underground where it is a serious environmental problem.

Even after the extraction process is complete, more energy is burned and more pollution is produced to make Tar Sands into something useful. Very high temperatures (up to 4,000 degrees Fahrenheit) and the addition of hydrogen are necessary to make synthetic oil from bitumen. Refineries processing Tar Sands tend to spew more sulfur dioxide pollution per barrel of production than non–Tar Sands refineries partly because of the much higher sulfur content of the bitumen material as compared to most unrefined petroleum.

Sulfur dioxide pollution increases health risks especially for those already suffering from lung or heart problems. And much of that pollution occurs in refinery fence-line communities that have higher levels of poverty and persons from minority groups than other communities.

Bitumen blends from Canada's Tar Sands are much more corrosive than conventional unrefined petroleum. That means, in addition to generating more toxic air pollution, refineries using Tar Sands have higher risks for pipeline breaks and spills. These spills can contaminate rivers and groundwater that are necessary for community health.

References

Grant, Jennifer Grant, Dan Woynillowicz, and Simon Dyer. "Clearing the Air on Oil Sands Myths." Calgary: Pembina Institute (June 2009).

Kelly, Erin N., et al. "Oil Sands Development Contributes Polycyclic Aromatic Compounds to the Athabasca River and its Tributaries." *PNAS* 107, no. 37 (2009): 16178–83.

Standing, Tom. "Canadian Oil Sands Misses Unrealistic Projection—Issues Another." *Energy Bulletin* (2009). http://www.energybulletin.net/node/50971. Accessed February 14, 2012.

Tenenbaum, David J. "Oil Sands Development: A Health Risk Worth Taking?" *Environmental Health Perspectives* 117, no. 4 (2009): A150–56.

Aaron Sanger is currently director of U.S. campaigns with ForestEthics, which means that he guides the U.S. campaign against Canada's Tar Sands as well as campaigns to protect forests threatened by corporate paper consumption and greenwash. After becoming dissatisfied with the limits to how he could support social change as a lawyer, he decided to become a full-time environmental campaign organizer. He went through the year-long Green Corps training for environmental organizers and ended up with ForestEthics, where his first job was leading a campaign to protect endangered forests in Chile.

IS THERE A PEAK OIL?

Mike Lynch

People have long feared that they would use up the Earth's resources, whether mineral or energy, and thought that population growth would overwhelm agricultural resources. Usually, these fears have arisen in response to temporary problems, such as drought or depletion of an important mine, but especially in the past few decades, academic studies have purported to show that resource scarcity was imminent. Petroleum has been a primary focus of many studies because of its importance and finite nature.

The finite nature of petroleum is not actually relevant. If it were, whale oil would still be in use. Rather, the size of the oil resource relative to the amount being used is what matters.

Many early oil industry experts worried that the resource was limited, even though the science was immature. Indeed, it was not until the 1901 discovery at Spindletop in Texas, the first American gusher, that the railroad industry, over 40 years after the initial oil well was drilled, decided there were sufficient supplies for them to safely switch from using coal as fuel.

But fears about depletion have continued to arise from time to time. President Jimmy Carter famously declared in 1977 that "the oil and gas we rely on is simply running out," but the price collapse in 1986 made many think that he had exaggerated the problem. Indeed, he clearly misunderstood that shortages of natural gas had been caused by price controls, not scarcity, as decontrol led to a boom in supply and lower prices.

But new arguments arose in the 1990s, as a series of studies by geologists argued that better data allowed for greater precision in resource estimation, and indicated that nearly half of the world's petroleum resource had been consumed, suggesting production would peak. They argued that this research was reliable because the field size estimates were made by geologists, and the mathematical techniques were highly reliable.

In fact, these geologists proved to be novices at statistical analysis, and made numerous mistakes. The assertion that the field size estimates were stable proved erroneous, and the predictions of production peaks for various countries proved repeatedly premature. Again and again, global production has passed the supposedly scientifically forecast peak.

Nearly all of this work is based on one outstanding mistake: that production trends must continue, whether for a field or a country, once they are declining. This assumes, often implicitly, that production levels are driven only by geology, and cannot be altered by human decisions, such as changes in taxes or the use of new technology to improve recovery. This is a fatal flaw.

Globally, about 35 percent of the oil in the ground is recovered with current technology, depending on the type of rock, chemistry of the oil (some is thicker, for example), and the investments made. A century ago, the amount recovered was only 10 percent, and advances in both geological knowledge and technology have increased it, in some cases, to 70 percent. Longer term, this recovery factor will increase, adding to the world's reserves, but most peak oil advocates argue that this is untrue, that essentially all technologies currently in use are decades old. The only reason for such an old assertion is that it allows the earlier claim, that field size estimates don't increase, to hold true.

Other mistakes abound, including the claim that geology causes oil production in a given field, country, region, and the entire planet should resemble a bell curve, when in fact, it rarely does. Or that peak production occurs when half of a field, country, region, and the planet has been produced, when such has rarely happened. And the argument that geologists are on one side of the debate (as believers in imminent peak oil) and economists on the other (as optimists about supply) is blatantly false, as few geologists or petroleum engineers subscribe to peak oil theories.

Perhaps the most damning evidence is the manner in which peak oil experts have claimed that most analysts agree with their estimate of two trillion barrels of recoverable oil, which was true until a decade ago, when new estimates appeared that put the amount at three and a half to four trillion barrels. No peak oil advocate has, to my knowledge, acknowledged this.

Discoveries of petroleum dropped sharply after 1973 because the Middle Eastern countries where most of the huge fields were found reduced exploration, because higher prices caused oil demand to slow sharply so that production had to be cut back. Especially after the second oil crisis, in 1979, OPEC members reduced production from 30 million barrels per day to 15 million barrels per day. Having a great abundance of discovered oil, countries such as Saudi Arabia and Iran simply ceased exploration.

More recently, when the oil price collapsed in 1998, most oil companies cut back drilling to save money, and some shifted funds to resource-rich areas of the former Soviet Union. As a result, production in the rest of the world stagnated or fell, which reinforced the mistaken perception that world oil production had peaked. the U.S. invasion of Iraq and mass dismissals of employees of the Venezuelan oil company meant that production in both countries dropped sharply, which was mistakenly thought to be permanent.

These problems are, however, transitory in nature and most of them are slowly being overcome. Iraqi production has surpassed pre-invasion levels and is growing rapidly. New discoveries have been made in areas such as India, East Africa, and deepwater regions off the United States, Brazil, and West Africa that are going to add significant production over the next decade. Additionally, oil companies have developed techniques that allow them to produce oil from shale, where the low porosity means that the oil only flows after the rock has been fractured.

Michael Lynch is president of Strategic Energy & Economic Research, Inc. He has worked primarily in academia and has served as president of the U.S. Association for Energy Economists, is a lecturer at Vienna University, and blogs on energy matters for US News and World Report.

WAVE ENERGY

Ana Brito e Melo

Ocean waves as well as ocean currents, tidal range (rise and fall), tidal currents, ocean thermal energy, salinity gradients, submarine hydrothermal vents, marine biomass, and offshore wind are all different energy resources that can be harnessed in the ocean to produce electricity, drinking water, heat, hydrogen, or biofuels.

The International Energy Agency's Ocean Energy Systems Implementing Agreement set up its vision for the international deployment of ocean energy:

- Worldwide, there is the potential to develop 748 GW of ocean energy by 2050.
- By 2030, the ocean energy deployment could create 160,000 direct jobs.
- By 2050 this level of ocean energy deployment could save up to 5.7 billion tons of carbon dioxide.

Ocean waves are created by the action of wind passing over the surface of the ocean. A good energy resource is found at latitudes between 30° and 60° (northern and southern hemisphere). Attractive areas for wave energy exploitation can be found particularly in the European Atlantic Arc, in the west coast of South America, in southeast Australia, and in coast of New Zealand.

Wave energy is a topic that has always raised curiosity and still stimulates the imagination of inventors and engineers. There are more than 1,000 ideas for wave energy devices patented in Japan, North America, and Europe (Falcão 2010). Currently, around 60 different ongoing developments are progressing. The diversity of concepts that are being proposed is due to the possibility of harnessing wave energy at distinct locations (in the coastline, in intermediate depths, and offshore), and with several approaches. Devices can be floating, fully submerged, or integrated in fixed structures. Furthermore, different energy conversion methods can be utilized, such as air turbines, water turbines, high-pressure oil-driven hydraulic motors, or direct linear generators. Usually the wind power industry is mentioned in this context to highlight that, while a quick convergence to a unique turbine design occurred in this sector, in wave energy there is no leading technology.

Wave energy exploitation is in a demonstration phase. Leading developers are testing full-scale prototypes and designing their first wave-energy farms. In a less advanced stage, there is a wider range of technologies being developed that have already achieved significant progress in the maritime environment in protected areas. In an immediately preceding stage a

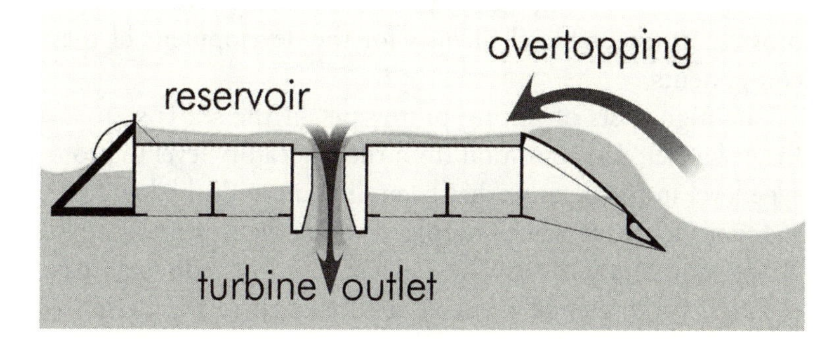

This drawing illustrates the overtopping principles and how the captured water is drained through the turbines. (Wave Dragon ApS)

vast group of technologies is being tested in laboratories. Many other technologies still at the conceptual phase may prove to be promising. The progress of a technology crucially depends on the ability of the developer to secure both private and public funding.

In recent years, a rapid development with new actors on scene has been noted, including large industrial companies and utilities. Only through such partnerships it will be possible to mobilize the necessary resources to advance to a pre-commercial phase.

A large proportion of the world's wave energy deployments is being done in the United Kingdom due to the government support for the industry in addition to the large attractive resource. Examples of UK support initiatives are the European Marine Energy Centre (for testing devices in the Orkneys (Scotland), the Wavehub in Cornwall for the deployment of the first wave farms, the Marine Renewables Proving Fund aiming to accelerate the most promising technologies towards the point where they qualify for the Government's Marine Renewables Deployment Fund, and the awarded leases from the Crown Estate for up to 1.6 GW (wave and tidal) at 11 UK sea sites.

The utilization of wave energy involves the development of control systems, reliable communications systems, mooring systems, transmission systems, and operation and maintenance strategies. The emerging status of wave energy technologies provides considerable challenges for the development of these components.

The high costs of testing prototypes in the sea is seen as a major barrier. The transition from the laboratory level to prototype level in the open sea has been difficult with the low funding levels available when compared with other energy sectors.

The development of wave devices requires reducing costs, through fundamental changes in the engineering design of devices. This also means more efficient use of materials and improvement of installation, operation, and maintenance procedures. The devices need to withstand extreme forces in storm conditions, and must be able to operate successfully in the harsh conditions of the marine environment. The UK Carbon Trust's Marine Energy Accelerator project recently concluded that "there is considerable scope for improving the fundamental energy performance of wave devices as experience is gained of how the devices operate in real-sea conditions. Improvements are possible through changes to the design of the device itself, specifically better coupling with the sea, and also from changes to the layout of devices in arrays."

Furthermore, there are significant opportunities for knowledge transfer from other sectors, such as offshore engineering. The ORECCA Project published in September 2011 the European Offshore Renewable Energy Roadmap, a combined roadmap developed for the offshore wind, wave, and tidal stream energy sectors, focused on the synergies, opportunities, and barriers to development. The report point out that co-location of devices represents an important opportunity with benefits from joint utilization of electrical infrastructure, operation and management teams, and vessels.

The development of offshore wave energy farms requires a reflection on how to order the ocean space that is subject to multiple traditional uses (fishing, shipping, aquaculture, submarine cables and natural and archeological reserves, among others). Marine renewable energy is seen as a newcomer and the necessity of early discussions about synergetic uses has been highlighted (Neumann 2010). There is actually a growing interest in the synergetic use of the ocean space. Several European projects are investigating multipurpose floating platforms for renewable energy and other uses (transport, aquaculture, and research and leisure purposes), such as the European funded TROPOS project initiated in February 2012 with a consortium of 18 partners, running for three years.

References

Falcão, A. "Wave Energy Utilization: A Review of the Technologies." *Renewable and Sustainable Energy Reviews* 14 (2010): 899–918.

Neumann, F. "Ocean Energy as Ocean Space Use—Only Conflicts or Also Synergies?" 2010. http://www.ocean-energy-systems.org/ocean_energy/in_depth_articles/synergies/. Accessed February 14, 2012.

Ana Brito e Melo has been working in marine renewable energy since 1993. She initiated her research activities at The School of Engineering (IST) in Lisbon and concluded a PhD in mechanical engineering in 2000, in the field of wave energy technologies. In 2004 she integrated the team of the newly founded nonprofit association—the Wave Energy Centre—and has been responsible for the coordination work of R&D activities and services. She has been also managing the secretariat of the International Energy Agency's Implementing Agreement on Ocean Energy Systems since 2002. In 2012 she was appointed executive director of the Wave Energy Centre.

THE SOLAR SOLUTION

Noah Davis

You may think of solar energy as a relatively recent player on the energy scene, with vast arrays of shining solar electric panels converting sunlight directly into electricity for use by modern appliances. However, since ancient times the sun's abundant energy has been used to heat homes, grow crops, grind grain, and move ships across the sea. In fact, virtually every other form of useful energy on the Earth can be traced back to its origins in sunlight. Solar energy is not just cutting-edge alternative energy—it is also the basis for the many forms of energy used by the ancient, modern, and future people of this planet.

Fossil fuels are actually stored sunlight, created millions of years ago when rainforests and algae in the sea harvested the sun's energy through photosynthesis. Vast quantities of these plant products later became trapped in the earth as coal and oil. So fossil fuels are a form of solar power (although these finite resources are nonrenewable and generate greenhouse gases when burned). Fossil fuels are just ancient biomass plant products grown by the sun.

Wind is caused by the sun's uneven heating of our planet. Energy is distributed from the equator, where the sun's rays are most intense, toward the North and South Poles through the constant movement of storms and winds. Wave energy is also powered by the wind. The water cycle, which powers hydroelectric dams, large and small, is fueled by the heat of the sun evaporating water molecules.

So rather than asking "what is solar energy?" the pertinent question becomes "what is not solar energy?" Nuclear, geothermal, and tidal energy are not powered by the sun (unless you dig deep for the connection). Everything else can be quickly traced back to solar energy from sunlight.

The term *alternative energy* can hardly be applied to solar at all, especially in a historical context. Humankind evolved over tens of thousands of years using only renewable energy from the sun. Then, over the past three centuries, we burned fossil fuels at an incredible rate, creating a society based on unprecedented growth and almost incomprehensible energy waste. Due to dwindling resources and increased pressure on the environment, this era will end soon. Our brief period of fossil fuel binging will return to a sustainable, renewable, familiar pattern—harvesting energy from the sun on a day-to-day basis in its many forms.

Today, solar energy is once again gaining attention as an abundant, distributed (available almost anywhere without powerlines or pipelines), inexhaustible means of powering just about anything. Architects are returning to designing homes that heat themselves simply by letting sunlight deep into the living space in winter. Those homes are often built of strawbales, a biomass product of the sun that is usually burned as waste. These simple, relatively low-tech improvements will go a long way toward returning the Earth's energy balance.

Solar hot water is also seeing regrowth. This relatively simple, effective technology pumps water or a heat exchange fluid through sunlight collectors that directly heat the fluid whenever the sun shines. In the United States in the 1970s and early 1980s, solar hot water was the leading solar technology on homes. In fact, the White House had a large system installed, and effective financial incentives led to large numbers of system installations on homes throughout the United States. Then, in the mid-1980s things changed: a new president was elected, the solar panels came down from the White House, financial incentives were removed, and the industry crashed. Today the technology has improved and so have the incentives—solar hot water is now the most effective and financially rewarding renewable energy system available for many homes.

At last we come to the technology that may have popped into your head when you first thought of solar energy: solar electric panels. Using the amazing, reliable, and effective process called photovoltaics, these technological wonders pump out electricity whenever sunlight strikes them. There are no moving parts, and nothing gets used up in the process.

Compare photovoltaic (PV) technology to electricity coming from other sources: coal-fired, natural gas–fired, hydroelectric, nuclear, and wind. Each one of those sources turns a generator to make electricity, a process that involves mechanical complexity, worn-out parts, and significant losses in efficiency. A *solar module* (which is the technical term for a solar electric panel) just sits silently in the sun and cranks out electricity, literally for decades.

It is no wonder that PV is getting serious attention around the world these days. It works. It looks sexy on a roof or in a field. And increasingly, it's an affordable way to make electricity. This last point is becoming very important. Financial analysts, energy executives, leaders of nations, and savvy homeowners are all honing in on one obvious fact: PV is on its way to becoming the cheapest form of generating electricity.

It has not been easy for PV to become affordable. Government incentives have been required. In the United States, you currently get back 30 percent of your investment for a home

PV (or solar hot water) system from the federal government. In Germany, the government is willing to pay you even more—about four times the going rate for electricity generated by your PV array. What does the German government get out of this? Their investment has made them world leaders in PV technology as the industry expands massively worldwide. Germany installed almost 25,000,000,000 watts of PV in 2011, more than the United States has installed in its entire history.

PV is on a path that will quickly make it competitive with other, dirtier forms of electrical generation. When that happens government incentives will no longer be necessary to draw people into installing solar. Eventually, as the industry achieves larger economies of scale, PV could become the cheapest form of electrical generation on the planet. So get used to seeing PV modules, they will soon be visible everywhere the sun shines.

Solar energy fueled the creation of human society, and it will continue to do so long after fossil fuels are abandoned. It takes many forms in many places and it is our responsibility to make effective use of this free, abundant, and limitless resource.

Noah Davis is manager of the Solar in the Schools program for Solar Energy International. He has an extensive background in newspaper, magazine, freelance, and technical writing.

4 Profiles

This chapter contains brief sketches of individuals and organizations who have or are actively involved in issues related to energy exploration, production, and consumption. The essays range from individuals with historical significance (such as James Watts and John D. Rockefeller) to those who are active in energy-related issues today (such as Richard Heinberg and Charles Keeling), as well as organizations favorable to, opposed to, or neutral with regard to important energy issues of the day.

American Council on Renewable Energy
1600 K Street NW, Suite 700
Washington, DC 20006
Phone: (202) 393-0001
E-mail: houston@acore.org
http://www.acore.org/

The American Council on Renewable Energy (ACORE) is a trade association consisting of more than 500 companies, industry associations, utilities, end users, financial and educational institutions, professional service firms, nonprofit organizations, and government agencies who are involved in one way or another in the production, distribution, financing,

---

Solar panels create energy at a National Renewable Energy Laboratory (NREL) center on the outskirts of Boulder, Colorado. The mission of NREL is to develop renewable energy and energy efficiency technologies that are transferable to the private sector. (Photo by John Moore/Getty Images)

use, and other aspects of renewable energy. Examples of current members include: AMCREF Community Capital, LLC; Abengoa Solar, Inc.; Alternative Energy magazine; Bechtel Power Corporation; Boston Carbon; Cape Cod Community College; Clean Energy Group; Distributed Sun; Duke Energy; Evergreen Recycling; FuelCell Energy, Inc.; Green Mountain Energy Company; Interstate Renewable Energy Council; MPI Solar; Novozymes North America; POET Biorefining; Professional Engineers in California Government; Royal Danish Embassy (Washington, DC); The Center for Sustainable Solar Farming; Volvo Group North America; and Windustry. The organization's stated mission is "to mov[e] renewable energy into the mainstream of America's economy, ensuring the success of the renewable energy industry while helping to build a sustainable and independent energy future for the nation."

ACORE was founded in 2001 for the purpose of bringing together groups of all kinds interested in moving renewable energy into the mainstream of America's economy. The organization's current activities are centered around six major programs:

- The ACORE Leadership Council provides a mechanism by which experts in various fields are able to express their views on essential issues in the field of renewable energy. The council has produced a number of papers on such issues and provides speakers on a variety of related topics.
- The Biomass Coordinating Council works to increase the role of biomass energy in the nation's energy equation, thus reducing its dependence on foreign fossil fuels. The council sponsors working groups, webinars, regional roundtables, and other educational events for the dissemination and discussion of topics related to biomass energy.
- The REFIN Directory is a tool for connecting groups with financial, personnel, and other resources with companies in need of such resources. As an example, REFIN made

possible the financing of three wind energy projects by Citibank.

- ACORE Regional Outreach Program is based on the fact that energy project are seldom based on the geography of state boundaries, but generally extended over wider areas. The outreach program attempts to bring together information about energy issues across larger regions (such as the West, Midwest, and Southeast) to facilitate the development and use of renewable energy resources.
- U.S. Partnership for Renewable Energy Financing (US PREF) is a program of experts from financial institutions, technology companies, project development companies, law firms, and other organizations to discuss policies for the advancement of renewable energy development in the United States.
- U.S.–China Program is a joint project between ACORE and Chinese groups, such as the Chinese Renewable Energy Industries Association (CREIA) for the fostering of cooperation between agencies and companies in the two countries with regard to renewable energy projects.

The primary mechanism through which ACORE works toward its goals are annual conferences that focus on three topics: marketing, finance, and policy. The RETECH marketing conferences bring together business leaders, investors, renewable company representatives, state and federal government officials, college and university educators, and exhibitors to discuss and learn about advances in wind, solar, geothermal, biomass, and other renewable energy technologies and supporting technologies. The 2010 RETECH conference included more than 3,000 attendees. The finance conferences provide executives of renewable energy companies with information about the ways to obtain financing for their projects and to make sure that those projects reach a successful conclusion. The policy conferences deal with all aspects of the production, storage, transmission, and consumption of all forms of renewable energy. The 2011 conference in Washington, D.C., for example, dealt with topics such as the effect of the current economic slowdown on the renewable energy industry, international growth in the field with related effects on competition by American industry, increasing demands by the Department of Defense for renewable energy in its activities, and possible effects of the expiration of biofuels incentives at the end of the year. The November 2011 U.S.–China workshop on renewable energy was held in Beijing on November 1, 2011. Topics included ways in which American companies can learn more about foreign investment in renewable energy based on the very successful Chinese experience, the very successful Chinese solar market and ways in which American companies may be able to partner with Chinese companies in this field, and the broader picture of renewable energy development in China and its implications for the U.S. industry.

ACORE annually publishes a variety of articles, reports, white papers, and other publications dealing with all aspects of renewable energy. Some examples include the article "Renewable Energy Trends 2010," for the magazine *Infrastructure Solutions*; the biannual report "Renewable Energy in America: Markets, Economic Development and Policy in the 50 States"; the slide collection "Renewable Energy Landscape: Market & Technology Overview"; the report "Renewable Energy Landscape: Market & Technology Overview"; and the conference report, "RETECH 2009 Summary Report." Of particular interest and value to the general reader is ACORE's compilation of statistics on the renewable energy industry, which is available as a slide show online. The slide show is available free to members of the organization and for sale to nonmembers. A sample version of the product is also available online at http://www.acore.org/publications/renewable-energy-landscape.

ACORE's website also provides a somewhat abbreviated introduction to various types of renewable energy, such as biofuel, biomass, geothermal, hydroelectric power, hydrogen energy, ocean and tidal energy, solar energy, waste to energy, and wind power. The website also has a useful and extensive section on careers in renewable energy.

American Petroleum Institute
1220 L St., NW
Washington, DC 20005-4070
Phone: (202) 682-8000
E-mail: http://www.api.org/contactus.cfm
http://www.api.org/

The American Petroleum Institute (API) is a trade advocacy association consisting of more than 400 companies involved in the production, refining, and distribution of petroleum, as well as support functions for the industry. The association was founded on March 20, 1919, after the end of World War I. During the war, a number of independent petroleum companies had worked together and with the federal government to produce and supply the petroleum products needed to win the war. When the war was over, the petroleum companies recognized the benefits of having a single trade organization that could represent their interests before the government and the general public. Such an association also had the potential for establishing, promulgating, and enforcing industry-wide standards for the production and the distribution of petroleum and petroleum products. The first of those standards, was announced in 1925. Today, the API administers a complex and a comprehensive set of more than 500 such standards covering all aspects of the petroleum and natural gas industry. The API currently has State Petroleum Councils in 33 states, all of which are located east of the Rocky Mountains.

API's activities can be broadly classified into one of four fields: standards, education and certification, statistics, and public advocacy and lobbying. Consistent with its early focus, the organization produces new and revised standards on a regular basis, covering virtually every conceivable aspect of petroleum and natural gas production and distribution. These standards cover topics such as the quality and uses of motor oil; dimensions of rings, plugs, threads, and other devices used in the industry; design and manufacture of vessels; measuring instruments and methods of calibration; safety and fire regulations; and inspection standards and procedures for refining operations. Certain API standards are well known both in and outside the industry as, for example, the API gravity number, which identifies the density of a petroleum product and the API number, which is a unique identifier for every well and exploration site in the United States.

The core of the API's training and certification effort includes programs such as: the API Monogram Program, which allows companies to apply for certification for the equipment they produce; APIQR (American Petroleum Institute Quality Registrar), which allows companies to have their products certified as approved for use in the industry by the API; ICP (Individual Certification Program), which provides individuals with training in and certification for certain types of specific skills used in the industry; witnessing programs, which train individuals in the skills required to observe and approve a variety of procedures used in the industry; and energy risk management, in which individuals are trained to assess and deal with the variety of risks that arise during the exploration for and production and distribution of petroleum and petroleum products. In addition to the training of workers in the field of petroleum and natural gas, API produces a variety of materials for the education of the general public. One of the best known of these resources is a website called Classroom Energy (http://www.classroom-energy.org/), at which the organization presents its own view on a number of energy-related issues, such as climate change, environmental protection, and oil spills. Critics have noted that these presentations are somewhat biased towards the view of the industry, which is hardly surprising or unreasonable given that advocacy is a legitimate function of the organization and its materials.

The API is particularly and justifiably proud of its work in producing statistics about every aspect of the petroleum and natural gas industries. It began this activity very early in its history, releasing weekly data on the production of crude oil as early as 1920. Among the core data sets produced by the organization are the *Weekly Statistical Bulletin*, *Monthly Statistical Report*, *Quarterly Well Completion Report*, *Imports and Exports of Crude Oil and Petroleum Products*, *Joint Association Survey on Drilling Costs*, and *Inventories of Natural Gas Liquids and Liquefied Refinery Gases*.

Finally, the API is a strong advocate for certain industry-favorable positions on a variety of energy, environment, political, and other issues. It exercises this function in a variety of ways, such as television advertising and news releases to the media. It also expresses its views through a number of lobbyists (16 in 2009) that attempt to influence legislation on issues related to petroleum and natural gas. In 2009 its budget for public advocacy was reported to be $3.6 million. Some of its positions on important national issues include support for additional oil and gas drilling as a way to increase the number of jobs available for Americans and improve the overall economy, for an extension of tax benefits to the oil and gas industry as a way of encouraging further exploration and improving the industry's efficiency, and for voluntary industry regulation as a way of dealing with environmental issues related to oil and gas production and distribution.

As do most trade organizations, the API sponsors and participates in a number of meetings, conferences, workshops, and other sessions in which new information can be disseminated and participants can interact with each other. Some of the topics of such meetings are exploration, production, distribution, offshore wells, storage tanks, oil and gas tankers, facility security, regulations, standards, and training and certification. The API also sponsors a prodigious publication program, producing more than 200,000 publications each year. Its online publications catalog includes sections on exploration and production, health and environmental issues, marine issues, marketing, refining, safety and fire issues, storage tanks, and petroleum measurement standards. As of early 2012, arguably the largest and most visible of API's political action efforts is an organization called Energy Citizens, a 450,000-person-strong group whose purpose is to represent industry viewpoints on a variety of national issues, such as offshore drilling, hydraulic fracturing, tax breaks for the petroleum industry, and exploitation of Canadian tar sands.

Association for the Study of Peak Oil & Gas International
Klintvagen 42
SE-756 55 Uppsala, Sweden
Phone: +46 471 76 43
E-mail: mikael.hook@fysast.uu.se
http://www.peakoil.net/

The Association for the Study of Peak Oil & Gas International (ASPO International) was formed by Colin J. Campbell in December 2000, largely as a result of a speech he gave on problems associated with the depletion of petroleum production. Campbell was then a petroleum geologist who had worked for a number of energy companies, including Texaco, British Petroleum, Amoco, Shenandoah Oil, Norsk Hydro, and Fina. He had also taught at Oxford University and had worked as an advisor to the Bulgarian and Swedish governments. He is currently retired and living in Cork, Ireland.

Campbell began writing about peak oil in the late 1990s, publishing a number of articles and one important book, *The Coming Oil Crisis* (Multi-Science, 1997) on the topic. His article on peak oil in *Scientific American* in March 1998, "The End of Cheap Oil" (written with French petroleum engineer, Jean H. Laherrère), is sometimes credited with convincing the International Energy Agency of the reality of a crisis in petroleum production and changing that agency's outlook on the future of world supplies of oil. His 2000 lecture at the Technical

University of Clausthal, in Germany, prompted Campbell to consider the possibility of forming a consortium of scientists working in the field of petroleum exploration to consider the implications of peak oil for world economies. Largely as a result of Campbell's initiative, a group of such individuals came together in 2002 at the University of Uppsala, Sweden, for the first International Workshop on Oil Depletion. The organizer of that conference, Kjell Aleklett, is now president of ASPO. In addition to Aleklett, the current board of directors of ASPO consists of members from Australia, China, Ireland, Italy, South Africa, Spain, Sweden, and the United States.

The association lists three main objectives in its mission statement: clarifying and evaluating the world's reserves of petroleum and natural gas; developing models for the path of oil and gas depletion, and assessing the economic, political, and technological implications of this model; and educating governments and the general public about the potential consequences of decreasing supplies of petroleum and natural gas.

Perhaps the most important of the organization's many activities is its annual conference on fossil fuel depletion. The ninth such conference was held in Brussels, Belgium, in April 2011. For many years, Colin Campbell also wrote a monthly newsletter for the organization. His last newsletter in that series, the 100th such publication, was released in April 2009, with Campbell's retirement from active involvement with APSO. For a short period of time, Campbell's newsletter was supplemented by a "President's Newsletter." Today, that newsletter has been replaced by a "Presidential Blog," which is available online at http://aleklett.wordpress.com/test/. The organization's website also has an invaluable collection of books; journal, magazine, and newspaper articles; academic theses; reports; statistical reviews; and Internet resources on all aspects of the peak oil and gas issue.

The formation of APSO in 2000 was followed soon thereafter by the creation of national organizations created for the same purpose and along the same general administrative lines as the parent organization. Today, national peak oil and gas organizations also exist in Argentina, Australia, Belgium, Canada, China, France, Germany, Hong Kong, Ireland, Israel, Italy, Kuwait, Mexico, Netherlands, New Zealand, Portugal, South Africa, South Korea, Spain, Switzerland, Sweden, the United Kingdom, and the United States.

ASPO-USA was founded in 2005 as a nonprofit 501(c) 3 organization to promote more efficient use of energy resources, encourage the transformation of communities to better deal with a post-oil future, and assist in the development of cooperative initiatives for an age of declining petroleum reserves and production. It has sponsored annual conferences on peak oil issues every year since its establishment. The 2011 conference was held in Washington, D.C. The association publishes a daily *Peak Oil News* and a weekly *Peak Oil Review,* both of which are gold mines of up-to-date news on every aspect of issues related to peak oil. These publications are both available on the ASPO-USA website (http://www.aspousa.org/index.php/newsletters/peak-oil-review/file-library/?dl_cat=2 and http://www.aspousa.org/index.php/category/peak-oil-review/peak-oil-review/). The website also has a very useful section introducing the basic concepts of peak oil, including topics such as peak oil basics, evidence for the timing of peak oil, peak oil data, unconventional liquids, commentaries on the International Energy Agency and U.S. Energy Information Administration assessments of peak oil, and mitigation strategies.

The ASPO affiliate in the United Kingdom is the Oil Depletion Analysis Centre (ODAC), an international public charity whose goal is to raise public awareness about problems associated with diminishing reserves and production of petroleum. It was founded in June 2001 to monitor and analyze information about petroleum supplies worldwide, to facilitate the exchange of information about this topic among interested parties, and to disseminate that information to governments, the general public, corporations, and stakeholders. On its web page, ODAC provides access to a number of important reports and articles about peak oil. It also provides a weekly electronic newsletter about developments related to petroleum exploration, discovery, production, and consumption at http://www.odac-info.org/newsletter. The website also includes a useful general introduction to the topic of peak oil, with a review of some major issues associated with that event.

Other national associations focus on more local issues, often in more limited ways. The South Africa ASPO organization, for example, features a documentary video *Peak Oil & South Africa–Impacts and Mitigation*, along with "News" and "Features" pages and a library of useful resources on peak oil.

Association for the Study of Peak Oil & Gas—USA. See Association for the Study of Peak Oil & Gas International

William Hart (1797–1865)

Historians often date the origin of the natural gas industry in the United States to the discovery of oil at Titusville, Pennsylvania, by "Colonel" Edwin Drake in 1859. While it is true that Drake's well produced gas as well as oil, this attribution is not correct. Indeed, natural gas had been discovered and put to use more than three decades earlier in an area relatively close Titusville: Fredonia, New York. The person responsible for that accomplishment was a tinsmith by the name of William Hart.

William Aaron Hart was born in the town of Bark Hasted, Connecticut, in 1797. Essentially nothing is known about his early life, except that he moved to Fredonia in 1819, reputedly bringing with him nothing other than his rifle and a pack of clothing. Again, little is known about his time in Fredonia until about 1825, except that he apparently pursued his craft as a tinsmith in the town. At some point, he became interested in the possibility of capturing and putting to use a seep of natural gas present adjacent to Canadaway Creek running through the center of town. His approach was to dig a hole (using shovels!)

that eventually reached a depth of 27 feet to gain improved access to the natural gas. He then trapped the gas in a gasometer, originally his wife's laundry tube with a hole cut into the bottom. The gasometer provided not only a reservoir for the gas, but also a means by which it could be measured (hence the term *gasometer*). Gas from the well was then transported to nearby homes through a series of hollow wooden pipes whose junctions were sealed with pitch. By August of 1825, Hart's natural gas system was being used to heat two stores, two shops, and one mill in the vicinity of the well. Three months later, 36 gas lamps had been installed in the town. Hart's project had more than utilitarian value as residents and visitors to the town, who came from far and wide, were astounded at the spectacularly beautiful, as well as useful, scene produced by the combustion of pure natural gas from the ground.

Hart soon expanded his efforts, extending the original well to a depth of 50 feet in 1850. At that point, he was producing enough gas to light 200 lamps in the area. Realizing the potential commercial value of Hart's operation, a group of Fredonia residents incorporated the Fredonia Gas Light Company in 1828, the first commercial natural gas company (and now the oldest) in the United States. The company continued to expand and improve its natural gas system in 1858 by digging a second well to a depth of more than 200 feet. By this time, the method of delivery natural gas to customers had also been vastly improved. The original wood-and-pitch system was, as one might guess, woefully inadequate because of the ease with which gas escaped from the pipes. The system was eventually replaced by a new method of delivery consisting of lead pipes that carried gas from the well to stores, shops, mills, or other locations. The gas was then distributed throughout the site by means of finer pipes made of tin (of course, Hart's specialty), which delivered a very small flow of gas to a room equivalent in brightness to about two normal candles. Turning a gas lamp on and off was the simplest of operations. During the day, a plug was placed in the end of a tin pipe, and at night it was removed and the gas lighted.

The rest of William Hart's life is nearly as much of a mystery as are earlier years. He established a home at 50 Forest Place in Fredonia, where he abandoned his work with natural gas and became a nursery operator, best known for the roses he grew. He also operated an amusement park that included a hot-water spa that, some historians believe, may have been heated by the combustion of natural gas. At some point, Hart and his family moved to Buffalo, where census records later listed his occupation as merchant, and that of his adult son as gas furnisher. Hart died in Buffalo on August 9, 1865, leaving behind his wife, son, and one daughter.

An interesting addendum to the story of William Hart and the Fredonia Gas Light Company relates to the geological formation in which Hart originally found natural gas. That formation is known as the Marcellus Shale and extends thousands of square miles underground from mid-New York State into mid-Tennessee and from mid-Ohio to the Pennsylvania–Maryland border. Geologists believe that the formation may hold trillions of cubic feet of natural gas, enough to supply a significant fraction of the United States' energy needs for decades into the future. The most likely method for recovering that resource, however, is hydraulic fracturing, or hydrofracing, a technology about which environmentalists and residents of the area have significant concerns. Whether the risks posed by hydrofracing are with the benefits provided by the huge supplies of natural gas is a question still being debated in the early 21st century.

Richard Heinberg (1950–)

Richard Heinberg is an educator, a journalist, and an environmentalist with special expertise and interest in the field of peak oil. He has written 10 books on the subject and has spoken throughout the United States and around the world on peak oil and associated topics, such as global climate change, food and agriculture, community resilience, and the current economic crisis. In 2006 he was honored with the M. King Hubbert Award for Excellence in Energy Education. In 2006,

Heinberg was awarded his master's degree in leadership and humanities from New College of California, where he also taught from 1998 to 2007 (when the college closed). At New College, he taught a course in Culture, Ecology, and Sustainable Community. In addition to his writing and speaking engagements, Heinberg is an illustrator and a book designer, and he continues to play the violin professionally. He is currently senior fellow at the Post Carbon Institute.

Richard Heinberg was born in Kirksville, Missouri, on October 21, 1950. After living in a number of locations in Missouri, Illinois, and Iowa, he entered the University of Iowa to study art and music (violin). His college career ended when the university was closed down during the Vietnam War as a result of anti-war protests in which, he says, he was active. He then decided to pursue a career in music. He taught himself to play the electric guitar, and worked with various rock bands for seven years before deciding that rock music would not be an acceptable long-term career choice. Heinberg then lived in a series of intentional communities, planned residential communities designed for groups of individuals with common social, political, environmental, religious, or spiritual philosophies. He met his wife, Janet Barocco, a master gardener, at an intentional community in southern California, after which they moved to their own home in Santa Rosa. By this time, Heinberg had achieved enough success to make his living as a professional writer.

Heinberg published his first book, *Memories and Visions of Paradise: Exploring the Universal Myth of a Lost Golden Age,* in 1989. The book summarized his thoughts on world mythology, a topic about which he had been reading and thinking for a decade. An expanded and revised edition of the book was published in 1995. In 1992, Heinberg began writing a monthly publication called *Museletter,* in which he shares his thoughts on a wide variety of topics that include "geopolitics, energy depletion, civilization and its unintended consequences, economics from a contrarian perspective, and suggestions for how to weather the coming energy and economic transition." Issue 230 of *Museletter,* published in July 2011, covered three topics dealing with depletion of energy resources and the language that government and industry use to talk about these issues. The newsletter is now available in PDF format online at http://richardheinberg.com/museletter.

Over the years, the focus of Heinberg's books has evolved from spiritual to energy topics. He published *Celebrate the Solstice: Honoring the Earth's Seasonal Rhythms through Festival and Ceremony* in 1993 and *A New Covenant with Nature: Notes on the End of Civilization and the Renewal of Culture* in 1996, before turning his attention to *The Party's Over: Oil, War and the Fate of Industrial Societies* in 2003, *Powerdown: Options and Actions for a Post-Carbon World* in 2004, *The Oil Depletion Protocol: A Plan to Avert Oil Wars, Terrorism, and Economic Collapse* in 2006, and *Peak Everything: Waking Up to the Century of Declines* in 2007. His latest book, *End of Growth: Adapting to Our New Economic Reality,* was published in 2011. Heinberg's books have been translated into eight languages, including Chinese, Japanese, Spanish, Portugese, and Italian. His articles have appeared in magazines and journals around the world, including *Alternative Press Review, The American Prospect, Canadian Dimension, Earth Island Journal, Ecologist, European Business Review, The Futurist, Pacific Ecologist, The Proceedings of the Canadian Association of the Club of Rome, Public Policy Research, Quarterly Review, Resurgence, Wild Matters,* and *Z Magazine.*

Heinberg has been guest speaker at more than three dozen colleges and universities across the United States and in a half dozen foreign countries. He has also spoken widely before a variety of professional organizations, including the Center for Strategic and International Studies in Washington, D.C.; the National School of Government in London; Lawrence Berkeley National Laboratories; the Development Bank of Southern Africa; the Canadian National Farmers Union; the Colombian Society of Engineers; the Association for the Study of Peak Oil annual workshops in Lisbon and Pisa; the Young Presidents

Organization of Guatemala; the Midwest Renewable Energy Fair; the annual Foundation for the Economics of Sustainability conference in Ireland; and the American Solar Energy Society. Heinberg has also appeared in a number of energy-related documentaries, such as *The End of Suburbia* (2004), *The Power of Community: How Cuba Survived Peak Oil* (2006), *Ripe for Change* (PBS; 2006), *Oil, Smoke & Mirrors* (2006), *Crude Impact* (2006), *What a Way to Go* (2007), *Escape from Suburbia* (2007), *Oil Apocalypse* (History Channel: Megadisaster series; 2007), *11th Hour* (produced and narrated by Leonardo DiCaprio, 2007), *Blind Spot* (2008), *A Farm for the Future* (BBC, 2009), and *Earth2100* (ABC Television, 2009).

M. King Hubbert (1903–1989)

Marion King Hubbert was a research geophysicist who spent most of his professional career at the Shell Research Laboratories in Houston, Texas. He is best known for his suggestion that oil production tends to follow a bell-shaped curve, in which production levels are very low at first, rise to a maximum level, and then begin to decrease. At the tail-end of the curve, oil production essentially drops to nearly zero. In his original paper on this topic, Hubbert predicted that oil production in the United States would peak in the 1970s, a prediction that later turned out to be correct. He later predicted that oil production worldwide would peak in the 1990s, after which production would decline relatively rapidly. For a variety of reasons, that prediction was wrong in terms of the date of peak production, although it probably is not incorrect in its overall concept.

Hubbert was born in San Saba, Texas, on October 5, 1903. His parents were William Bee and Cora (née) Hubbert. The Hubbert family had come to the San Saba region of Texas in the 1840s from France, and William and Cora remained in that region for most of their lives. When King was four years old, his mother organized a one-room school for neighborhood children, an act that impressed the young boy. He later said that a one-room school was a good way to earn an education as students were not separated from each other, and everyone "had to work together."

After graduating from high school in San Saba, Hubbert enrolled at Weatherford Junior College, which remains in existence today as the oldest community college in the state. He then expressed an interest in continuing his studies at the University of Chicago. Although his mother approved of the idea, his father was reluctant to pay the cost of a college education. Hubbert decided, therefore, to earn enough during the summer to finance at least his first year at Chicago. He became a part of the group of men (most of them illegal migrants) who traveled from field to field in boxcars. By the end of the summer, Hubbert had earned enough to enroll at Chicago. The university refused to accept his Weatherford credits, however, and he had to begin his studies as a first-year student. When the university demanded that Hubbert declare a major, he was unable to make up his mind among his favorite topics. So he declared a triple major, in geology, mathematics, and physics. It was the first triple major in the university's history.

In a 2006 interview, Hubbert's nephew, Michael Hubbert, relates an interesting story from his Chicago days. At one point in a geology class, he was introduced to Darcy's Law, a statement of the way a fluid flows through a porous medium. Hubbert decided that the law was correct, but that the interpretation being offered in class was incorrect. When he made this point to the instructor, he was ridiculed, although, over time, his position was found to be correct.

Hubbert was awarded his BS degree in 1926 and his MS degree in 1928. He then enrolled in a doctoral program at Chicago, from which he received his PhD in 1937. During the decade between his master's and doctoral degrees, he held a number of positions. He worked as an assistant geologist for the Amerada Petroleum Company, at the Illinois State Geological Survey, and the U.S. Geological Survey. He also accepted a position teaching geology at Columbia University in 1930, a post he held until 1940. During World War II, Hubbert also worked as a senior analyst at the Board of Economic Warfare in Washington, D.C.

In 1943, Hubbert joined the research staff at the Shell Oil Company in Houston, Texas. He eventually became director of the research laboratory. He retired from Shell in 1964 and took a position as senior research geophysicist at the U.S. Geological Survey, a post he held until 1976. Hubbert died on October 11, 1989.

Although Hubbert is best known for his prediction of peak oil, he made a number of other contributions in the field of petroleum geology. Perhaps as a result of his interest in Darcy's Law from his Chicago days, Hubbert carried out extensive research on the movement of petroleum through rock and soils. He also attempted to quantify the petroleum reserves available in the United States and other parts of the world, research that eventually led to his conclusions about the limitations of petroleum reserves. Hubbert also provided a mathematical demonstration of a principle long suspected among geologists, namely that rocky material in the Earth's crust is plastic, capable of deforming over long periods of time.

For a number of years, Hubbert was interested in a political philosophy known as technocracy, a philosophy that government functions can best be performed by technocrats and scientists who are familiar with the technical details of most problems. In 1933, he was a cofounder of a group advocating this philosophy—Technocracy, Incorporated (TI)—a group that remains in existence today. Hubbert's affiliation with TI was, in fact, the reason he was terminated in 1943 from his position at the Board of Economic Warfare, which regarded TI as politically suspect and perhaps dangerous to the war effort.

Hubbert was elected to the National Academy of Sciences in 1955 and the American Academy of Arts and Sciences in 1957. He served as president of the Geological Society of America in 1962. Among his many honors and awards were the Arthur L. Day Medal (1954) and the Penrose Medal (1973) of the Geological Society of America, the Rockefeller Public Service Award of Princeton University (1977), the William Smith Medal of the Geological Society of London (1978), and the Elliott Cresson Medal of the Franklin Institute (1981).

Intergovernmental Panel on Climate Change
c/o World Meteorological Organization
7bis Avenue de la Paix
C.P. 2300
CH-1211 Geneva 2
Switzerland
Phone: +41-22-730-8208/54/84
Fax: +41-22-730-8025/13
E-mail: ipcc-sec@wmo.int
http://www.ipcc.ch/

The Intergovernmental Panel on Climate Change (IPCC) was created in 1988 as a joint program of the United Nations Environment Program and the World Meteorological Organization for the purpose of measuring and assessing possible global climate change. The organization was established by United Nations General Assembly resolution 43/53, adopted on December 6, 1988, with the charge of initiating action:

> leading, as soon as possible, to a comprehensive review and recommendations with respect to:
>
> (a) The state of knowledge of the science of climate and climatic change;
>
> (b) Programmes and studies on the social and economic impact of climate change, including global warming;
>
> (c) Possible response strategies to delay, limit or mitigate the impact of adverse climate change;

(d) The identification and possible strengthening of relevant existing international legal instruments having a bearing on climate;

(e) Elements for inclusion in a possible future international convention on climate.

The organization's original charge has been modified and expanded in later documents, as outlined in the document "Principles Governing IPCC Work," adopted at the 14th general session in Vienna in October 1998, the 21st general session in Vienna in November 2003, and the 25th general session in Mauritius in April 2006.

In 1990, the IPCC issued its first report as directed by its founding document, a report known as the First Assessment Report. Among the conclusions presented in that report was a statement that the natural greenhouse effect in the Earth's atmosphere was being augmented by anthropogenic contributions, most importantly, gases emitted during the combustion of fossil fuels. Authors of the report that current levels of greenhouse gases could be stabilized only by the immediate reduction by 60 percent of the release of greenhouse gases by human activities. They also predicted a rise in the Earth's annual average temperature of about 0.3°C (0.5°F) per decade (with an uncertainty range of 0.2 to 0.5°C [0.4 to 0.9°F] per decade) during the 21st century. They concluded with some projections as to possible effects that might result from changes in the Earth's climate of this magnitude.

IPCC has published three additional assessment reports, in 1995, 2001, and 2007. A fifth assessment report is scheduled for release in 2013 or 2014. Each report provides conclusions developed by three working groups that focus on different aspects of the climate change problem. For example, in the Second Assessment Report, Working Group I focused on updating scientific and technical information about greenhouse gases, changes in global temperatures, and related issues. Working Group II considered the feasibility of various technical and economic fixes for global climate change. And Working Group III analyzed long-term social and economic consequences of global climate change of the magnitude anticipated by the technical experts. The Second Assessment Report was used by attendees at the Kyoto conference on climate change in 1997 as the latest and most authoritative description of the science, technology, and economic and social impact of global climate change.

The third and fourth assessments followed the pattern of the earlier reports, discussing updated scientific and technical information about climate change, as well as projections as to possible environmental, economic, social, and other consequences of this change. Shortly after the fourth assessment was issued in January 2007, the IPCC was awarded a share of the 2007 Nobel Peace Prize (along with former U.S. Vice President Al Gore Jr.) "for their efforts to build up and disseminate greater knowledge about man-made climate change, and to lay the foundations for the measures that are needed to counteract such change."

The IPCC consists of an administrative staff of about 10 individuals who plan, coordinate, and oversee the work of the organization. That work is actually carried out by large groups of scientists organized into one of three working groups, as described above, working out of centers in the United States, Germany, and Japan. The 2007 report, for example, was written by 152 lead authors along with 498 contributing authors, based on a review of more than 6,000 peer-reviewed studies, with reviews by 625 scientific experts and 26 review editors from 40 different countries.

The Fifth Assessment Report will place greater emphasis on potential social consequences of global climate change, with more information on the effects in individual regions and the relationship of other weather events, such as El Niño and monsoon activity. More attention will also be paid to clarifying the uncertainties involved in making some of these estimates and projections and the development of new scenarios resulting from global climate change.

In addition to the core assessment reports, IPCC also publishes a number of other reports and documents related to global climate change, such as *Carbon Dioxide Capture and Storage, Safeguarding the Ozone Layer and the Global Climate System: Issues Related to Hydrofluorocarbons and Perfluorocarbons, Methodological and Technological Issues in Technology Transfer, Emissions Scenarios, Land Use, Land-Use Change, and Forestry, Aviation and the Global Atmosphere,* and *The Regional Impacts of Climate Change: An Assessment of Vulnerability.*

IPCC also sponsors a scholarship program that provides funding for young men and women interested in various aspects of the climate change issue. In 2011, the organization announced that it had awarded nine such scholarships for research in fields such as the underlying science of climate change (students from Madagascar and Mozambique), impacts of climate change on aquatic ecosystems, water availability, health and agriculture (students from Nepal, Burkina Faso, and Benin), climate-related disaster management (a student from Uganda), climate modeling and assessment of the impacts of climate change (a student from Ethiopia), and adaptation and mitigation options for different sectors and assessment of socioeconomic implications (students from Bangladesh and Tanzania).

International Energy Agency
9, rue de la Fédération
75739 Paris Cedex 15
France
Phone: +33 1 40 57 65 54
Fax: +33 1 40 57 65 59
E-mail: info@iea.org
http://www.iea.org/

The International Energy Agency (IEA) was founded in 1974 in response to the world crisis in energy supplies resulting from the oil embargo announced by the Organization of Arab Petroleum Exporting Countries. The embargo had been established as a way by which Arab states could object to support of the state of Israel by the United States and other Western nations. Confronted with the sudden and drastic loss of its most important energy source, petroleum, a number of nations banded together to form the IEA. IEA's initial goals were to develop more sophisticated and accurate statistics on known and estimate petroleum reserves and oil production and consumption and to investigate ways of ameliorating the effects of the dramatic loss in petroleum supplies for developing nations. Since 1974, IEA's mission has expanded somewhat and now includes four major themes.

The first of these themes is energy security, which involves the promotion of diversity and flexibility of energy supplies in order to avoid the worst consequences of fuel crises like that of the 1974 oil embargo. The second theme is environmental protection, which recognizes the deleterious effects of extensive fossil fuel use. A recent feature of this theme is a greater emphasis on the possible global climate effects resulting from the combustion of coal, oil, and natural gas. The third theme is economic growth, which emphasizes the importance of developing energy policies that ensure the continued economic development of all nations, whether developed or developing. The fourth theme is engagement worldwide, which is a specific acknowledgment that the energy and environmental policies facing nations of the world transcend national borders and can only be solved by international cooperation.

Only member states of the Organisation for Economic Cooperation and Development (OECD) are permitted to belong to the IEA. Current members are Australia, Austria, Belgium, Canada, Czech Republic, Denmark, Finland, France, Germany, Greece, Hungary, Ireland, Italy, Japan, Luxembourg, Netherlands, New Zealand, Norway, Poland, Portugal, Slovakia, South Korea, Spain, Sweden, Switzerland, Turkey, the United Kingdom, and the United States. Although prohibited from inviting non-OECD nations to join the agency, the IEA has a well-developed program for working with such nations through its

Directorate of Global Energy Dialogue (GED). Established in 1993, the GED attempts to better understand the status of energy systems in non-OECD nations and to share with those nations some of the knowledge and expertise developed through IEA programs. The agency currently has bilateral and regional agreements with a number of non-OECD nations and regions to achieve these goals, including Brazil, Caspian and Central Asia, Central and Eastern Europe, China, India, Mexico, the Middle East, Russia, Southeast Asia, Ukraine, and Venezuela.

The primary decision-making body of the IEA is the Governing Board, which consists of the energy ministers of all member states, or their designated representatives. Decisions made by the Governing Board are carried out by the Secretariat, which consists of researchers and other scholars. The Secretariat's work involves the collection of data on energy supplies, production, and consumption; sponsorship of conferences and other meetings on energy issues; assessment of energy conditions in member states; development of projections for global energy futures; and proposals for national energy policies in member states.

The work of the Secretariat is organized around a number of specific topics, for which a variety of reports and publications are generally available. The current topics of interest to the IEA include coal, carbon dioxide capture and storage, cleaner fossil fuels, climate change, electricity, energy efficiency, energy indicators, energy policy, energy predictions, energy statistics, fusion power, greenhouse gases, natural gas, oil, renewable energy, sustainable development, and technology. For the general public and experts in the field, IEA's most important contributions may well be their regular reports on a variety of important energy-related issues. In the area of statistics, the agency publishes a host of reports on a monthly, quarterly, and annual basis. Perhaps best known of these is its annual *Key World Energy Statistics,* published regularly for the past 10 years. Other statistics publications deal with energy prices and taxes, energy statistics for OECD and non-OECD states, oil information, natural gas information, renewables information, and carbon dioxide emissions from fuel combustion. The agency also publishes a quarterly publication, *Oil, Gas, Coal, and Electricity.*

Other fields for which publications are available are global energy dialogue, which includes publications on energy issues in non-OECD nations; energy and environment, which deals with topics such as the environmental and climate effects of fossil fuel combustion; renewable energy, which considers technical, social, economic, and political aspects of the development of renewable energy sources; energy efficiency, which discusses ways in which conservation can extend the lifetime of existing fossil fuel resources; energy technology network, which reports on agreements made among member states to improve their energy production and consumption patterns; policy analysis and cooperation, which reviews energy policies, practices, and projections for specific member states; and energy technology, which reviews proposed and in-place changes in technology to make more efficient use of existing and proposed energy sources. An especially interesting feature of the agency's website is its "Fast Facts" section, which lists tidbits of information about many aspects of energy. Each item is cited from an IEA publication, to which a link is provided in each case.

In addition to its print publications, the IEA sponsors a number of conferences and other meetings, provides speakers for a variety of professional and community events, and testifies before governmental agencies in member nations. In 2011, for example, the agency sponsored, cosponsored, or participated in meetings on energy outlooks, in Saudi Arabia; energy storage, in Paris; renewable energy, in New Delhi; electric vehicles, in Shanghai; a clean energy symposium, in Astana, Kazakhstan; energy efficiency in buildings, in Brussels; and low-income weatherization projects, in Dublin.

Charles Keeling (1928–2005)

Charles Keeling was an atmospheric chemist who is probably best known for collecting a set of data about the amount of carbon dioxide in the atmosphere over a period of more than 40 years. When Keeling's data are plotted as a graph, they show a gradual increase in the concentration of carbon dioxide in the atmosphere from his earliest measurements in 1958 (315 parts per million) to their present level in mid-2011 (392 parts per million). The graph also shows fluctuations within each year that reflect the growth and the death of land planets and, hence, the amount of carbon dioxide removed from the atmosphere as the result of photosynthesis. Keeling's work was of enormous influence in the field of global climate change because it provided the first quantitative data about the amounts of carbon dioxide in the atmosphere, historical changes in that metric, and a variety of factors that affect it.

Charles David Keeling was born in Scranton, Pennsylvania, on April 20, 1928. He attended the University of Illinois, from which he received his bachelor's degree in 1948, and Northwestern University, where he earned his PhD in 1954. Keeling then moved to the West Coast where he was a postdoctoral fellow at the California Institute of Technology (CalTech). In 1956 he accepted a position at the Scripps Institution of Technology, where he remained for the rest of his academic career. He was appointed professor of oceanography at Scripps in 1968.

Keeling became interested in measuring the atmospheric concentration of carbon dioxide early in his career. While he was still a postdoctoral student at CalTech, he built a device for making this measurement and spent three weeks with his young family at the Big Sur State Park in California taking measurements. Before long, he had become convinced that atmospheric concentrations of carbon dioxide had increased significantly over those indicated by studies of air bubbles trapped in ice masses dating back thousands of years. He decided to obtain more precise and more extensive data on this point.

Keeling was aided in the advancement of this line of research by two factors: the beginning of the International Geophysical Year (IGY) in 1958, and the assistance of his colleague and advisor, Roger Revelle, one of the founders of the IGY. Keeling was able to obtain an IGY grant to study atmospheric carbon dioxide atop Mauno Loa, on the Big Island of Hawaii, more than 10,000 feet (3,000 meters) above sea level. Keeling chose this location to avoid confounding factors associated with forested lands, urban or industrial areas, or other locations where the concentration of carbon dioxide might be significantly different from that in the pure atmosphere itself. An observatory remains atop Mauna Loa, where data about changing carbon dioxide concentrations are still collected on a regular basis.

Keeling's work involved more than simple data collection. He also developed methods for determining the source of carbon dioxide (natural vs. anthropogenic), its potential fate (an ocean or land-based sink vs. storage in the atmosphere), and the relative amount of carbon dioxide resulting from anthropogenic versus natural sources (a relatively constant 55%/45% ratio). Although the implications of Keeling's research now seem obvious, they were not always so to other scientists or to funding agencies. The National Science Foundation discontinued funding of Keeling's research in January 1964 suggesting that the work he was doing was too routine to justify further funding. So many complaints were heard, however, that the agency renewed Keeling's grant four months later, and data collection began once again. (The four-month data gap in the Keeling curve during which Keeling's measuring devices were shut down is, however, a vivid reminder of the almost-end of this classic project.)

Keeling's research has produced what is arguably the single most important set of data regarding changes in atmospheric concentration of carbon dioxide in history. The relationship between those data and possible global climate change remains a point of some (but increasingly less) debate among specialists in the field. Keeling himself tended to avoid speculating on the long-term climatic, social, political, economic, environmental, and other implications of a growing level of atmospheric carbon dioxide. He eventually did become much more interested, however, in developing models of the potential climatic effects of the changes he was observing, another field to which he made important contributions.

Keeling had a number of interests beyond the field of science. He was a talented pianist and had, at one time, considered a career as a concert pianist. He was also a founder and founding director of the University of California, San Diego, Madrigal Singers. Among his many honors are the Second Half Century Award of the American Meteorological Society, the Tyler Prize for Environmental Achievement, the Maurice Ewing Medal of the American Geophysical Union, the Blue Planet Prize of the Science Council of Japan and the Asahi Foundation, and the U.S. National Medal of Science. He was a Guggenheim fellow at the Meteorological Institute at the University of Stockholm from 1961 to 1962 and visiting professor at the University of Heidelberg (1969–1970) and the University of Bern (1979–1980). Keeling was a fellow of the American Academy of Arts and Sciences, the American Geophysical Union, and the American Association for the Advancement of Science, and a member of the National Academy of Sciences. He died of a heart attack on June 22, 2005, at his summer home in Hamilton, Montana.

John L. Lewis (1880–1969)

John L. Lewis was perhaps the greatest spokesperson and fighter for the rights and welfare of coal miners in the history of the United States. He served for 40 years, from 1920 to 1960, as president of the United Mine Workers of America (UMWA). In 1935, Lewis also joined with leaders of seven other unions to form the Congress for Industrial Organization (later the Congress of Industrial Organizations [CIO]), eventually to become the largest union organization in the United States. He was chosen first president of the CIO, a post he held until 1940.

Many coal miners of the late 19th and early 20th century lived lives of almost unbelievable destitution and danger (a statement that is not necessarily restricted to the United

States or that time period). They often lived in towns owned by the coal company, sent their children to schools operated by the company, purchased everything they needed at company stores, and lived out their lives in constant and endless debt to the company. They worked in conditions designed to produce maximum profit for companies with limited attention to the safety of workers. Accidents were common and often devastating. The three largest mining accident in American history, for example, all occurred within a six-year period between 1907 and 1913, when 362 workers were killed in an explosion at the Fairmont Coal Company in Monoghan, West Virginia; 263 workers at the Stag Canon No. 2 mine in Dawson, New Mexico; and 259 workers in a fire at the Cherry mine in Cherry, Illinois. It is not an exaggeration to say that miners often owed their lives to the company, in almost every respect. John L. Lewis spent his life trying to change that reality.

John Llewellyn Lewis was born in Cleveland, Iowa, on February 12, 1880, to Thomas H. and Ann Watkins Lewis, both immigrants from Wales. Thomas Lewis was a miner, who was also active in the union movement, a reason enough for him to be blacklisted by most mining companies. As a result, the family, including young John's seven siblings, moved frequently while Thomas looked for employment. During one period of relative stability the family lived in Des Moines where John was able to complete elementary school and three years of high school. However, he then left school to take a job as a miner and field worker in Lucas, Iowa, where he soon became active in union activities. In 1907, he attempted to start his own business as a feed-and-grain distributor, but soon failed in that endeavor. His first political effort, a campaign for mayor of Lucas, also failed. As a consequence, he returned to mining and to labor activities.

Lewis did not remain in the miners for long. He soon moved to Panama, Illinois, where he began to devote all of his time to union activities. He was elected president of the UMWA there in 1909 and two years later was chosen as a full-time labor organizer by Samuel Gomphers, president of the American Federation of Labor (AFL). He rapidly worked his way up through the union becoming statistician, vice president, and then acting president of the organization in 1919. A year later he was elected president, a post he held until 1960.

Lewis soon became known as a fiery orator, who maintained strong-handed control of the union. He worked to centralize many of the programs and activities of the union, converting it from a loose confederation of local groups to a powerful national unit with far more influence over mining activities. Although a lifelong Republican, Lewis gradually began to throw his support to the Democratic Party. In the elections of 1928 and 1932, he publicly supported Herbert Hoover, but privately provided union financial support to Democratic candidate Franklin D. Roosevelt in 1932. He did so because he believed that the Democratic Party was more committed to the kind of mining reforms for which he and his union stood. As a reward for his support, Roosevelt appointed Lewis to the Labor Advisory Board and the National Labor Board of the National Recovery Administration in 1933, providing him with a platform from which he could push for many of the mining reforms advocated by the UMWA.

With the support of the Roosevelt administration, the union movement began to grow to new heights and influence. Lewis was convinced that it could improve even further by expanding its reach to groups of workers previously untouched by the movement. To that end, he and leaders of nine other large unions formed the CIO in 1935, a subgroup within the AFL. As the CIO began to organize across a number of trades, including steel, rubber, automobile production, meat packing, and electrical equipments, it began to outshadow its parent group, the AFL. In the dispute that developed between the two giant organizations, the AFL expelled the CIO in 1938, a split that was not to be healed until the formation of the American Federation of Labor–Congress of Industrial Organizations (AFL-CIO) was formed in 1955. Lewis was chosen as the first president of the CIO and was so successful that, by 1937, his subgroup of the AFL had more members than did the parent.

Lewis' professional biography provides a checkered story. He undoubtedly made significant progress in the lives of individual miners (and other workers), resulting in better wages and working conditions. He was also instrumental in the passage of the first really effective piece of mine safety legislation, the Federal Mine Safety Act of 1952. Although revised and updated, that act continues to serve as the cornerstone of the nation's mine safety program. At the same time, Lewis remained in autocratic control of his union, packing its offices with his favorites, and keeping some locals and divisions under his personal control. Only with the passage of the Labor Management Reporting and Disclosure Act (the so-called Landrum-Griffin Act, after its Congressional sponsors) in 1959 did Lewis sense that his time as union leader had passed and he resigned as president of the UMWA in 1960. He then retired to his long-time home in Alexandria, Virginia, where he died on June 11, 1969.

Amory Lovins (1947–)

Amory Lovins is a well-respected environmental scientist perhaps best known for his development of the concept of soft energy in the 1970s. The term *soft energy* refers to the emphasis on energy efficiency and the use of alternative, renewable fuels as an alternative to the world's current dependence on centralized production and distribution of fossil fuels.

Amory Bloch Lovins was born in Washington, D.C., on November 13, 1947. He grew up in nearby Silver Spring, Maryland, and in Amherst, Massachusetts. He entered Harvard College in 1964 but left after two years to enroll at Magdalen College, Oxford University, England. His decision to leave Harvard, he later explained, was a consequence of the college's unwillingness to allow him to continue his studies without declaring a major, which Lovins was not ready to do. At Magdalen, Lovins concentrated on physics, but did not complete his degree there either. Nonetheless, the college decided to hire

Lovins as a junior research fellow and don at the college in 1971. In order to accommodate his academic standing to his position at Magdalen, the college awarded him a master's degree by special resolution.

By the end of the academic year in 1971, Lovins decided not to continue his affiliation with Magdalen since his primary field of interest by that time—energy studies—was not yet recognized as a legitimate topic for doctoral study. Instead, he moved to London where he worked as a consultant to industrial and governmental agencies on energy-related issues. He returned to the United States in 1978, where he served as Regents' Lecturer in Energy and Resources at the University of California at Berkeley. In following years, he was also distinguished visiting scholar at the University of Oklahoma (1979), Regents' Lecturer in Economics at the University of California at Riverside, (1980), and member of the Energy Research Advisory Board of the U.S. Department of Energy (1980–1981).

In 1982, Lovins and his wife, Hunter, cofounded the Rocky Mountain Institute (RMI) in Snowmass, Colorado. Lovins has continued his affiliation with RMI ever since, currently serving as chairman and chief scientist at the institute. Although his primary professional association has been with RMI for three decades, Lovins has continued to write, speak, and teach in a variety of settings. He was Henry R. Luce Visiting Professor of Environmental Studies, Dartmouth College, in 1982; Distinguished Visiting Professor of Environmental Design, University of Colorado also in 1982; cofounder and director of E SOURCE, a consulting firm that helps companies make better use of energy supplies; Oikos Visiting Professor at the University of St. Gallen (Switzerland) Business School in 1999; cofounder and chairman of Hypercar, Inc. (now Fiberforge Corporation) in 1999; visiting lecturer at the College of Environmental Engineering, Peking University in 2002, and MAP/Ming visiting professor at the Stanford University School of Engineering in 2007.

Lovins has authored or coauthored 29 books, some of the best known of which are *World Energy Strategies: Facts, Issues, and Options; Nuclear Power: Technical Bases for Ethical Concern; Soft Energy Paths: Towards a Durable Peace; The Energy Controversy: Soft Path Questions and Answers; Non-Nuclear Futures: The Case for an Ethical Energy Strategy* (with John H. Price); *Least-Cost Energy: Solving the CO_2 Problem; Energy Unbound: A Fable for America's Future* (with L Hunter Lovins; Seth Zuckerman); *Consumer Guide to Home Energy Savings; Reinventing Electric Utilities: Competition, Citizen Action, and Clean Power; Small is Profitable: The Hidden Economic Benefits of Making Electrical Resources the Right Size;* and *Winning the Oil Endgame: Innovation for Profit, Jobs and Security.*

Lovins has been awarded honorary doctorates by 10 universities around the world. He has also received a number of notable prizes and awards including the Right Livelihood Award (the so-called Alternative Nobel Prize), Delphi Prize of the Onassis Foundation, Nissan Prize, Heinz Award for the Environment, Lindbergh Award, World Technology Award, Happold Medal of the U.K. Construction Industry Council, Shingo Prize, Benjamin Franklin Medal, Jean Meyer Award, Blue Planet Prize, and Volvo Environment Prize. In 1997, Lovins was named a MacArthur Fellow, and in 2000, he was named one of *Time* magazines "heroes of the planet."

National Renewable Energy Laboratory
1617 Cole Blvd.
Golden, CO 80401-3393
Phone: (303) 275-3000
http://www.nrel.gov/

The National Renewable Energy Laboratory (NREL) was established by the Solar Energy Research Development and Demonstration Act of 1974 as the Solar Energy Research Institute. The NREL began operations in July 1977 and was designated a national laboratory of the U.S. Department of

Energy (DOE) in September 1991. The NREL main campus is located in Golden, Colorado, and its National Wind Technology Center is located about 20 miles north of the main campus. The agency also has administrative offices in Washington, D.C.

NREL conducts research on all forms of renewable energy, including hydrogen fuel cell and related technologies; bioenergy; and wind, solar, and geothermal technologies. Some of the research units at which this research occurs are the National Center for Photovoltaics, the National Bioenergy Center, the Thermochemical Users Facility, the Battery Test Facility, the Thermal Conversion Facility, the Ethanol Process Development Unit, the Solar Radiation Research Laboratory, the Photovoltaics Outdoor Test Facility, the Solar Furnace, and the Renewable Fuels and Lubricants Research Laboratory, all located at the Golden campus. The agency's overall objective is to develop cleaner, more reliable, more affordable energy options that will reduce air pollution and greenhouse gas emissions, strengthen the nation's energy security, improve electric grid operations, boost local economic development, and increase energy and economic efficiency. Although the agency's primary emphasis is on America's energy equation, it also has a commitment to improving energy options throughout the world and has contributed to renewable energy projects in rural communities in Africa, Asia, and South America.

The work of the NREL falls into four general categories: energy analysis, science and technology, technology transfer, and applying technologies. The agency's energy analysis components involves the analysis of technical systems, markets, policies, and levels of sustainability as well as the development of models and tools for the study of renewable resources. These studies have produced a large number of publications, perhaps most useful of which for the general public are the annual *Renewable Energy Data Book* and *Renewable Energy Market* Data, both of which are available for download on the agency's website (http://www.nrel.gov/analysis/).

The NREL science and technology component includes research studies undertaken by the agency on the major forms of renewable energy and has resulted in reports on topics such as hydrogen gas turbine development, advanced vehicle drive systems, cellulose digestion in the production of bioenergy, data availability for home energy retrofits, biohybrid fuel cell development, materials development for improved solar cells, and analysis of current offshore wind energy developments. Examples of some of the specific research areas undertaken by NREL in just one area, advanced vehicles and fuels research, are electrical vehicle grid intergation, fleet test and evaluation, fuel technology impacts, petroleum based fuels, power electronics, nonpetroleum-based fuels, regulatory support, and vehicle systems analysis.

The technology transfer component of NREL's work involves cooperative efforts between the agency and private corporations for the implementation of research conducted at NREL. This cooperation involves work conducted jointly between the agency and private corporations, the licensing to corporations of discoveries and advances made at NREL, and work carried out by NREL for energy corporations. One of the success stories touted by NREL in its technology transfer program has been the development of a new type of material for use in the construction of solar cells. The new material makes use of a new kind of metal foil called rolling assisted biaxially textured substrates (RABiTS), originally developed at NREL on which silicon crystals are deposited out of the gas phase in a process developed by the Ampulse Corporation. NREL and Ampulse researchers hope that their new material will provide a breakthrough in the process by which solar cells can be produced with a significant reduction in the cost of production of cells.

The applying technology component of the NREL mission involves the agency's assistance to businesses and institutions, federal government agencies and facilities, state and local governmental agencies, tribal communities, and international and regional governments and institutions in implementing and adapting NREL discoveries and advances to each specific entities special needs and conditions. As an example, the NREL assumed a major role in the rebuilding of the town of Greensburg, Kansas, after the community had been essentially destroyed by a tornado that hit the area in May 2007. NREL was able to make suggestions for reconstruction that eventually reduced the community's carbon dioxide emissions by 36 percent over pre-tornado levels.

The NREL has an extensive collection of publications, many of which are available online. They include popular topics such as *Consumer's Guide: Get Your Power from the Sun* (brochure); *Elements of an Energy-Efficient House* (fact sheet); *Borrower's Guide to Financing Solar Energy Systems: A Federal Overview* (booklet); *Junior Solar Sprint: An Introduction to Building a Model Solar Car* (instructional booklet); *Wind Energy Benefits* (fact sheet); *Look Back at the U.S. Department of Energy's Aquatic Species Program: Biodiesel from Algae; Close-Out Report* (report); *Lessons Learned from Existing Biomass Power Plants* (booklet); *Axial Flux, Modular, Permanent-Magnet Generator with a Toroidal Winding for Wind Turbine Applications* (technical report); *International Performance Measurement and Verification Protocol: Concepts and Options for Determining Energy and Water Savings, Volume I* (protocol); and *Concentrating Solar Power: Energy from Mirrors* (PDF file).

NREL also conducts an energy leadership program called the Executive Energy Leadership Academy. The program is focused on providing individuals with a nontechnical explanation of the opportunities available for the use of renewable energy options in a variety of government and business settings. Two program options are available, one consisting of five two-day sessions completed over a period of five months, and the other consisting of one three-day session completed within a one-month period.

Oil Depletion Analysis Centre. See Association for the Study of Peak Oil & Gas International

Organization of the Petroleum Exporting Countries

Helferstorferstrasse 17
A-1010
Vienna, Austria
Phone: +43 1 21112 3302; +43 1 21112 3301
E-mail: See website at http://www.opec.org/opec_web/en/contact/356.htm
http://www.opec.org/opec_web/en/index.htm

The Organization of the Petroleum Exporting Countries (OPEC) was founded at the Baghdad Conference of September 10–14, 1960, by five petroleum-exporting nations: Iran, Iraq, Kuwait, Saudi Arabia, and Venezuela. The five founding members were later joined by Qatar (1961), Indonesia (1962), Socialist Peoples Libyan Arab Jamahiriya (Libya; 1962), United Arab Emirates (1967), Algeria (1969), Nigeria (1971), Ecuador (1973), Angola (2007), and Gabon (1975). Gabon was a member for two decades, from 1975 to 1994. Ecuador suspended its membership from December 1992 to October 2007, but is now a full-fledged member again. Indonesia suspended its membership in OPEC in January 2009 when it became a net petroleum-importing nation, although it indicated its intentions and desire to rejoin the organization if and when it once more becomes an oil-exporting nation. From 1960 to 1965, OPEC maintained its headquarters in Geneva, Switzerland, but moved that facility to Vienna, Austria, where it has remained ever since.

OPEC was organized at a time when many world powers, such as Great Britain, France, Germany, Belgium, and the Netherlands were divesting themselves from colonies in Africa and the Middle East over which they had maintained control for the better part of a century. At the time, petroleum development and production was controlled largely by the so-called Seven Sisters, seven major international oil companies who felt, in general, that they had little or no obligation to contribute to the development of the nations from which petroleum was being extracted. (The Seven Sisters were Standard Oil of New Jersey, Standard Oil Company of New York [now ExxonMobil], Standard Oil of California, Gulf Oil, Texaco [now Chevron], Royal Dutch Shell, and Anglo-Persian Oil Company [now BP]). At the time of OPEC's founding, the Seven Sisters controlled about 90 percent of the world's oil production. The formation of OPEC reflected a philosophy that new and developing nations had a right and obligation to control the development and use of their own national resources, including petroleum and other fossil fuel products.

OPEC adopted a set of three principles at the Baghdad Conference, principles that still serve as guidelines for the organization's activities today. The three principles called for cooperation among member nations to secure fair and stable oil prices for producing nations; an efficient and dependable supply of petroleum to consuming nations; and a fair return on capital for individuals and organizations who invest in the production and distribution of petroleum supplies. These principles were formalized in 1968 in the Declaratory Statement of Petroleum Policy in Member Countries, which stated that all countries have the inalienable right "to exercise permanent sovereignty over their natural resources in the interest of their national development."

As of early 2011, OPEC members collectively control about 80 percent of the world's crude oil reserves and about 45 percent of the world's crude oil production. Because of these resources, the organization maintains enormous political and economic control over nations that depend heavily on petroleum for their economic survival and development, which essentially includes almost every country in the world. An example of the impact OPEC can have on national policies can be seen in the 1973 Arab Oil Embargo, a name given to an action taken by Arab members of OPEC organized under the name of the Organization of Arab Petroleum Exporting Countries (OAPEC). The oil embargo occurred between October 1973 and March 1974 in response to actions taken by the United States in support of the national of Israel in the so-called Yom Kippur war that began on October 6, 1973. Two weeks after the war began, OAPEC members declared an embargo on petroleum shipments to any nation that sided with or aided Israel in the war. At first, the embargo applied only to the United States, although it was later extended to other nations, such as the Netherlands, South Africa, Rhodesia, and Portugal. The oil embargo came to an end on March 17, when OAPEC members (except for Libya) announced that they would renew shipments to all consuming nations, the United States included.

The oil embargo was the first, but not the only time, at which OPEC and other oil-exporting nations have used petroleum as an economic and political weapon. The effectiveness of that approach is seen in the changes in oil prices in countries affected by the embargo. The price of crude oil nearly quadrupled during the embargo, going from about $3 to about $12 a barrel (a modest price in today's terms, when crude oil sells at about $100 a barrel). During the same period, the price of gasoline in the United States rose from 38.5 cents a gallon to 55.1 cents a gallon (an increase of 43%).

In addition to its direct participation in the development, production, and distribution of petroleum, OPEC has been active in a variety of other fields, most important of which is probably the OPEC Fund for International Development (OFID). OFID was created at a 1975 meeting of heads of state of OPEC nations. The purpose of the organization is to provide assistance to developing nations who are trying to meet basic needs for their populations. OFID is funded by OPEC through loans and grants to both public and private institutions. Examples of the types of projects supported by OFID in recent years include road construction in a number of regions, including Albania, Bangladesh, and

Cameroon; energy development projects in Gambia, Kenya, and Paraguay; fisheries development in Mozambique; emergency, research, and technical assistance grants, usually in cooperation with various nongovernmental organizations; and special grants for the Palestine National Authority and for various HIV or AIDS groups. As of May 2011, OFID had made financial commitments by way of loans and grants in the amount of $13.168 billion, of which $8.045 billion had already been disbursed.

Post Carbon Institute
613 4th Street, Suite 208
Santa Rosa, CA 95404
Phone: (707) 823-8700
Fax: (866) 797-5820
E-mail: at http://www.postcarbon.org/contact/
http://www.postcarbon.org/

The Post Carbon Institute is a nonprofit 501(c)3 think tank that attempts to deal with an interconnected group of problems facing the world in the 21st century. These problems include scarcity of natural resources (including fossil fuels), global climate change, increasing disparity between the haves and have-nots of the world, and the challenge of developing a sustainable world economy. The institute collects information on such issues and makes it available to individuals, governmental agencies, communities, and businesses in an effort to help them develop plans that will reflect the economic and ecological realities of the future world.

The Post Carbon Institute had its origins in 2003 as a subsidiary program of the MetaFoundation. The MetaFoundation, in turn, had been created in 2000 as a means of contributing to the development of new programs and organizations whose goal was to find ways of dealing with new environmental challenges of the 21st century. The first such program created by the MetaFoundation was Global Public Media, an Internet broadcasting station designed to provide in-depth news and analysis of environmental issues for those interested in planning for the Earth's future. In 2003, both Global Public Media and the Post Carbon Institute were incorporated as affiliated initiatives of MetaFoundation. Over the past decade, the relationship among these entities has evolved so that today MetaFoundation is listed as the dba (doing business as) name of the Post Carbon Institute. The Post Carbon Institute was originally founded by Julian Darley and Celine Rich, who served as president and executive director, respectively, until their departure from the organization in 2009. The current executive director of the institute is Asher Miller.

Post Carbon Institute has assembled 28 fellows to provide expertise on the topics in which it is most interested. They include individuals such as Erika Allen (community food systems), Zenobia Barlow (ecological literacy), Hillary Brown (buildings and design), Majora Carter (social justice and communities), Richard Douthwaite (economics and money), David Fridley (renewable energy and biofuel), David Hughes (fossil fuels), John Kaufmann (government and peak oil), Bill McKibben (climate, ecology, and economy), David Orr (climate, education, and communities), William Rees (ecology and resilience), Brian Schwartz (health and peak oil), Tom Whipple (peak oil), and Peter Whybrow (culture and behavior).

The institute has four primary goals. The first is to build awareness and understanding of the challenges facing the world in the post-carbon world. The second is to foster collaboration among individuals and groups with common problems and challenges. The third is to integrate knowledge about environmental and social challenges, treating issues in a holistic, rather than piecemeal, fashion. The fourth is to inspire action, helping people and communities to find realistic solutions to the problems they face. The institute focuses its efforts on about a dozen major issues, such as: global climate change; the consumption of natural resources and waste management issues arising out of this consumption; the planning for and development of communities able to operate on a sustainable basis; the development of a philosophy and practices that take a more holistic approach to the role of humans in the general environment; the promotion of knowledge about and skills for dealing with 21st century issues; the development of new and better ways to produce energy for the world's growing demands; the developmental of alternative agricultural technologies that do not depend so completely on fossil fuel products; the re-invention of governmental agencies that will enable them to respond more effectively and efficiently to today's challenges; developing better ways of dealing with modern health problems; and finding ways of controlling world population growth.

The Post Carbon Institute's activities fall into four general categories, which it categorizes as "Setting the Agenda," "Changing the Conversation," "The Resilience Network," and "Supporting Transition." Setting the Agenda refers to a wide range of books, articles, reports, and op-eds that build awareness of current challenges, present alternative views about these challenges, and suggest new models for dealing with problems facing communities. Changing the Conversation consists of a variety of ways by which the institute can provide a wider introduction to its philosophy and programs, including a speakers bureau; mechanisms for interaction with the academic community; outreach to the general public through websites, social media, newsletters, briefings and exchanges, press outreach, collaborative efforts with nongovernmental organizations, and EnergyBulletin.net, an online collection of articles about all aspects of energy issues. The Resilience Network is an online cooperative effort of "the nation's most credible, influential people and organizations" with the goal of developing an integrated understanding of the most critical problems facing the world today. Supporting Transition refers to the financial and technical assistance provided by Post Carbon to Transition United States, a nonprofit organization whose goal it is to help individual communities build more locally focused sustainable economies.

Perhaps the best known publication of the institute is *Post Carbon Reader,* edited by Richard Heinberg and Daniel Lerch, and published by the University of California Press (2010). The book consists of 34 essays dealing with all aspects of the post-carbon world, with special emphasis on the arrival of peak oil and other peak resources. The organization's website also contains a very large collection of videos on all aspects of its works and the problems and challenges with which it is concerned. The website also provides a link to the institute's monthly e-newsletter, *PCI Newsletter*, at http://www.postcarbon.org/publications/newsletters/.

John D. Rockefeller (1839–1937)

John D. Rockefeller is perhaps the most famous oil magnate who has ever lived. In 1863, Rockefeller and his business partner, Maurice B. Clark, built a petroleum refinery in Cleveland, Ohio, for the purpose of extracting kerosene, the new fuel of choice for heating, lighting, and other purposes, from crude oil. Over the next few decades, he went on to build and buy additional refineries for the production of kerosene and gasoline, whose importance had skyrocketed because of the invention of the automobile. By 1880, Rockefeller's plants were refining 90 percent of the petroleum in the United States, and only a few years later, they were also refining 90 percent of the world's crude oil. In 1870, Rockefeller founded Standard Oil of Ohio as parent company for his enterprises. The company eventually grew to become the largest corporation in the world, and one of the most complicated industries in the United States. By 1880 Rockefeller had become a millionaire and was ultimately to be named the richest man in the world, the first person to have become a billionaire in terms of current dollar valuation.

John Davidson Rockefeller was born on July 8, 1839, in Richford, New York, to William Avery and Eliza Davidson Rockefeller. William Rockefeller led a somewhat checkered life, with a tendency to leave his wife and family and return to them on a somewhat irregular basis. He apparently took a second wife at some time in his life. He was known to his friends as "Big Bill" or "Devil Bill" Rockefeller. His wife was a patient and an understanding woman with strong religious beliefs, devoted to keeping her family of six children in tact. The family moved to Moravia, New York, in 1851, where young John attended Oswego Academy, and then to Cleveland, where he attended Cleveland Central High School. At the age of 16, John took a job as an assistant bookkeeper at Hewitt & Tuttle, a food wholesaler. Four years later, he formed his own produce firm with a friend, Maurice B. Clark. In 1863, Rockefeller, Clark, and three other partners built an oil refinery in Cleveland for the purpose of extracting kerosene from crude oil. At the time, the world's developing economies were slowly changing over from the use of whale oil for lamps to the less expensive kerosene that could be obtained from the distillation of petroleum.

Over the next decade, Rockefeller expanded his business interests both horizontally and vertically. The former effort involved buying up nearly all of his refining competitors in the Cleveland area, 22 of 26 businesses over a four-month period in 1872 alone. He then extended his reach to other American refineries until he had gained control of virtually all of his competitors. He also began to absorb other companies involved in one way or another with petroleum production and distribution. He built pipelines, rail lines, train cars, and delivery trucks with which Standard was able to ship kerosene directly from refineries to consumers' homes. At one point, he came into conflict with another giant corporation, the Pennsylvania Railroad, over a dispute about methods of shipping Standard products. In an attempt to fight Rockefeller on his own turf, the Pennsylvania began building its own refineries and distribution system, a battle it eventually lost when the company agreed to sell all of its petroleum-related business to Standard Oil in 1877.

The imbroglio between Standard Oil and the Pennsylvania Railroad brought to the attention of politicians and the general public the growing power of Rockefeller's empire. Individual states began passing anti-trust laws that limited the extent to which one corporation could control all or most of the market for a particular product. Rockefeller attempted to evade these laws with a number of sophisticated moves, including formation of the Standard Oil Trust, a corporate entity that brought together 41 of Rockefeller's individual companies and provided them (and him) with greater protection against attacks by individual states. This effort was eventually thwarted, however, when the Sherman Antitrust Act in 1890 forced Standard Oil to break up into a number of different entities. That process was completed in 1911 when the U.S. Supreme Court forced Standard of Ohio to divide its operations into 34 distinct companies, the two largest of which were the Standard Oil Company of New Jersey (Jersey Standard), which later become Exxon, and Standard Oil Company of New York (Socony), which later became Mobil.

On his retirement in 1897, at the age of 58, Rockefeller still held about a quarter of the stock in his 34 diverse companies, an asset that has been valued at about $900,000,000 as of the early 1920s. After his retirement, Rockefeller devoted most of his energies to philanthropic endeavors, for which he is widely remembered today. Among his many contributions were funding for a small college in Atlanta for African American women, named after his wife's family, Spelman College; grants to the University of Chicago, amounting to more than $80 million during his lifetime; founding of the Rockefeller Institute for Medical Research (now Rockefeller University) in New York City; establishment of the General Education Board, to study public education; creation of the Rockefeller Sanitary Commission for the Eradication of Hookworm disease; founding of the China Medical Board, to create a modern medical system for that country; establishment of Schools of Public Health at Harvard University and the University of Michigan; creation of the Berkshire Music Center at Tanglewood, Massachusetts; founding of the Peking Union Medical College in China; and funding for the establishment of departments of psychiatry at a number of colleges and universities, including Chicago, Duke, Harvard, McGill, Tulane, and Yale. By some estimates, Rockefeller spent about $550,000 on such philanthropic activities during his lifetime.

Rockefeller died of arteriosclerosis on May 23, 1937, at his home in Ormond Beach, Florida. He is buried in Lake View Cemetery in Cleveland. Rockefeller was survived by four daughters and one son, John D. Rockefeller Jr., who assumed the reigns of the Rockefeller Foundation after his father's death. A number of members of the Rockefeller family have become famous in the own right, including Nelson Aldrich Rockefeller, John's grandson, 49th governor of the state of New York, 41st vice president of the United States, and one-time candidate for the U.S. presidency; and John Davison "Jay" Rockefeller IV, 29th governor of the state of West Virginia, and senator from the state from 1984 to the present time.

Ida M. Tarbell (1857–1944)

Ida Tarbell was a teacher, an author, and a journalist who is best known for her investigative reporting on Standard Oil of Ohio and its chief executive, John D. Rockefeller, in the early 1900s. Her research and writing fall into a category generally known as muckraking journalism, although Tarbell herself rejected this characterization and said that she was simply writing a historical account of the company and its leader. Tarbell's book, *The History of the Standard Oil Company,* was listed fifth in a 1999 *New York Times* survey of the 100 most important books of 20th-century American journalism. The book first appeared as a series of 19 installments in *McClure's* magazine between 1902 and 1904. Tarbell's exposé of the previously unrecognized activities of Rockefeller, his fellow industrialists, Standard Oil, and similar large corporations was a staggering revelation to most Americans who had known almost nothing about the operation of big business in the United States. Her articles and books were credited by many people as providing a motivating force for the eventual breakup of Standard Oil and for passage of the Sherman Antitrust Act in 1890. Tarbell's articles and book were based on extensive interviews with anyone and everyone who had the most remote connection with Standard Oil and on studies of thousands of documents related to Rockefeller and his business activities. The one person with whom she never spoke was Rockefeller himself.

Ida Minerva Tarbell was born in the village of Hatch Hollow, Erie County, western Pennsylvania on November 5, 1857. Her parents were Franklin Summer and Esther Ann McCullough Tarbell, a joiner and a teacher, respectively. At the age of three, Ida's family moved to Titusville, Pennsylvania, the site at which oil was first discovered in the United States in 1859. Franklin Tarbell became interested in the oil business and formed his own company to produce and refine the product. In about 1872, Tarbell's business was nearly ruined by a scheme devised by John D. Rockefeller and a consortium of major railroads to give Rockefeller's oil companies deep discounts for the transport of their products. The scheme was discovered before it was ever put into practice, but the possibilities of its effects on small oil producers was devastating to much of the market.

Ida Tarbell graduated as the head of her class at Titusville High School, and then matriculated at Allegheny College, in Meadville, Pennsylvania, in 1876, where she was the only female student. She earned her bachelor of arts degree in biology in 1880, but ended up teaching a host of subjects, including four foreign languages, geology, and mathematics, at Poland Union Seminary in Poland, Ohio. At the end of two years, she decided that she no longer had an interest in teaching, and returned to Allegheny, where she was awarded her MS degree in 1883. (Allegheny later awarded her two honorary degrees, an LHD in 1909 and an LLD in 1915.) She then decided to try to make her way as a journalist, taking a job as a writer for *The Chautauquan,* an educational supplement designed for use in home-study courses that evolved out of the Chautauqua movement of the late 19th century. She was eventually promoted to managing editor of the publication. In 1890, Tarbell moved to Paris to do postgraduate studies at the Sorbonne and to continue her writing on a variety of topics. Her articles brought her to the attention of Samuel McClure, publisher of *McClure's* magazine, who offered her a job on the magazine staff. She accepted and continued writing for the magazine until 1906. After she left *McClure's,* Tarbell and fellow journalists from the magazine, Lincoln Steffens and Ray Stannard Baker bought *American Magazine,* which she coedited and wrote for until 1915. Tarbell then gave up active journalism to become a popular speaker on the Chautauqua lecture circuit for more than 20 years. She died at the age of 86 in Bridgeport, Connecticut, on January 6, 1944.

Tarbell's earliest triumphs at *McClure's* were two long series of articles on Napoleon Bonaparte and Abraham Lincoln, articles she later expanded to a number of books on the two men, including *A Short Life of Napoleon Bonaparte* (1895), *The Early Life of Abraham Lincoln* (1896), *A Life of Napoleon Bonaparte* (1901), *The Life of Abraham Lincoln* (1900), *Father Abraham* (1909), *In Lincoln's Chair* (1920), *Boy Scouts' Life of Lincoln* (1921), and *He Knew Lincoln* (1922). Tarbell also published her own autobiography in 1932, *All in the Day's Work.* Tarbell was inducted into the National Women's Hall of Fame, in Seneca, New York, in 2000. The U.S. Postal Service issued a stamp in her honor in 2002 as part of a series honoring women journalists. Tarbell's home in Easton, Connecticut, was declared a National Historic Landmark in 1993.

One of the great inconsistencies of Tarbell's life was her rejection of the fundamental principles of what was later to be called the Women's Rights Movement. In spite of her winning personal success, she believed that women were not physically, mentally, or emotionally suited to take an active part in business and politics. Their proper role, she said, was in the home, where they should serve as wives, mothers, and helpmeets. She opposed giving women the right to vote, arguing that they were not intellectually capable of making such important decisions, and she either declined to participate in or actively opposed a number of women's causes, such as the movement to extend family planning information to women. She said that women who sought a career outside the home were "freaks" and "misfits," doomed to failure and an unhappy life.

U.S. Department of Energy
1000 Independence Ave., SW
Washington, D. C. 20585
Phone: (202) 586-5000
Fax: (202) 586-4403
energy.gov
E-mail: The.Secretary@gq.doe.gov

Many agencies of the U.S. government have responsibility for one or another aspect of energy production, transportation, consumption, and environmental effects. The Mine Safety and Health Administration in the U.S. Department of Labor, for example, is charged with preventing death, disease, and injury from mining and, in general, promoting safe and healthful workplaces for the nation's miners. Similarly, the Nuclear Regulatory Commission is responsible for the overall safety of the nation's nuclear facilities. The one agency with the broadest and most comprehensive responsibility for energy matters in the United States, however, is the U.S. Department of Energy (DOE).

The DOE was established on August 4, 1977, when President Jimmy Carter signed the U.S. Department of Energy Organization Act, Public Law 95–91 (91 Stat. 565). The law was passed by the U.S. Congress largely because legislators felt that the United States had an inadequate federal mechanism for dealing with potentially devastating issues, such as the oil embargo invoked by the OPEC in 1973. Public Law 95–91 transferred a number of agencies with energy regulating authority to the new DOE, such as Federal Energy Administration, the Energy Research and Development Administration, the Federal Power Commission, along with divisions, bureaus, and units within other federal agencies. The law also created de novo a number of new divisions within the DOE, including an Energy Information Administration, Economic Regulatory Administration, Office of Inspector General, Office of Energy Research, Federal Energy Regulatory Commission, Leasing Liaison Committee, and Federal Energy Regulatory Commission. The department began operations on October 1, 1977, under the direction of the first Secretary of Energy, James R. Schlesinger, who had earlier served as Director of the Central Intelligence Agency under President Richard M. Nixon, and as Secretary of Defense under Nixon and President Gerald R. Ford.

The DOE currently consists of about a dozen separate offices with responsibility for various aspects of the nation's energy agenda. These offices include the Office of Civilian Radioactive Waste Management, Office of Electricity Delivery and Energy Reliability, Office of Energy Efficiency and Renewable Energy, Office of Environmental Management, Office of Fossil Energy, Office of Legacy Management, Office of Nuclear Energy, Science and Technology, Office of Science, National Nuclear Security Administration, Office of Secure Transportation, Office of Intelligence and Counterintelligence, Energy Information Administration, and Federal Energy Regulatory Commission. In additional to the topic-related divisions, four additional offices deal with large federal power-generating projects in various parts of the country, including the Bonneville Power Administration, in Portland, Oregon; Southeastern Power Administration, in Elberton, Georgia; Southwestern Power Administration, in Tulsa, Oklahoma; and Western Area Power Administration, in Lakewood, Colorado. In addition, the DOE operates a number of important research facilities, including the Lawrence Livermore National Laboratory, Lawrence Berkeley National Laboratory, Fermi National Accelerator Laboratory, Los Alamos National Laboratory, and National Renewable Energy Laboratory. The agency is also responsible for operations of the Strategic Petroleum Reserves operation in New Orleans, Louisiana.

The DOE has about 16,000 employees, with an additional 93,000 employees working under contract with the agency. In 2011, President Barack Obama included a request for more than $29.5 billion for the DOE in his fiscal year 2012 federal budget. The largest portion of that budget was for nuclear defense activities, such as weapons activities ($7.6 billion), environmental cleanup associated with weapons activities ($5.4 billion), defense nuclear nonproliferation ($2.5 billion), and naval nuclear reactors ($1.5 billion). Nuclear weapons activities account, overall, for about 60 percent of the DOE budget. The next largest item in the President's request was energy-related research, accounting for about $5.4 billion of his budget for the agency. The largest single increase in the budget (44%) was for research and development in the area of energy efficiency and renewable energy, $3.2 billion. The agency's budget for fossil fuel programs, primarily on research and development, was reduced in the President's budget request by a percentage almost equal to that of the alternative energy increase, 44 percent, to a total of about $520 million.

The DOE budget request for 2012 highlights the reality that a major portion of the agency's operations are still aimed at dealing with issues related to nuclear weapons research, development, regulation, and cleanup. However, that budget also illustrates the shift toward greater attention to nondefense energy issues of the 21st century, such as energy conservation, renewable energy, and the faster and more efficient commercialization of energy research findings by private companies. Some of the technologies it is now promoting in its search for new research proposals include advanced materials; biofuels and biomass; building energy efficiency; electricity transmission and distribution efficiency; energy storage; geothermal energy; hydrogen and other fuel cell development; hydropower, wave, and tidal technology; solar voltaic and solar thermal technologies; innovative vehicles and fuels; and wind power.

An interesting example of the new approach being tried by DOE for dealing with current energy issues is the concept of Energy Innovation Hubs. Energy Innovation Hubs are designed on the model of the great research laboratories developed during the Manhattan Project (e.g., the Los Alamos and Oak Ridge laboratories) for the development of nuclear weapon technology, and the AT&T Bell Laboratories, responsible for the development of so many important electronic advances in the past half century. The hubs will consist of researchers from a variety of disciplines from government, academia, and industry, focusing on some specific energy-related challenge, such as the development and distribution of solar-powered electricity. The first three hubs were established in 2010 with budgets of $25 million each. That number was doubled in Obama's 2012 budget request to six. The three original hubs focused on the development of fuels directly from sunlight (the Joint Center for Artificial Photosynthesis), improvement in nuclear reactor design and operation (Nuclear Energy Modeling an Simulation), and efficient energy retrofit of buildings (the Greater Philadelphia Innovation Cluster for Energy-Efficient Buildings). The three proposed new hubs will deal with improving batteries and energy storage systems, developing the best materials possible for clean energy production, and creating a smart and more efficient electricity grid.

James Watt (1736–1819)

James Watt is generally regarded as one of the founders of the Industrial Revolution because of his contributions to the development of the steam engine. He can not be said to have invented the device, since a number of more primitive engines operated by steam had been developed prior to his work on the invention. Indeed, the concept of a steam engine itself is relatively simple, and it is hardly surprising that a number of inventors prior to and during Watt's lifetime had created devices operating on the general principle. That principle is perhaps best illustrated by the invention on which Watt's own steam engine was based, the steam engine invented by Thomas Newcomen (1664–1729) in 1712. In Newcomen's engine, steam produced by a boiler was introduced into a cylinder fitted with a piston. The piston was attached to a rocking arm that was pushed downward with the introduction of steam into the cylinder. Immediately after steam had been introduced into the cylinder, a shot of cold water was also injected into the cylinder, causing the steam to condense. Conversion of the steam back into water resulted in the creation of a partial vacuum in the cylinder, allowing the rocking arm to return to its original position. The result on one cycle of the engine's operation, then, was a single up-and-down movement of the rocking arm and a consequent conversion of heat energy into the mechanical energy of the rocking arm. Watt's great improvement on the Newcomen engine was the realization that the squelching of steam by cold water during each repetition of the cycle resulted in the loss of large amounts of heat energy. His improved version of the engine consisted of two cylinders, one that was always kept warm and into which steam was introduced, and a second in which spent steam was allowed to condense. This version of the steam engine produced the same results as did the Newcomen version, but heat energy was not continuously being wasted in a condensation cycle, as was the case with the original design. Although the steam engine went through a number of modifications at later dates, Watt's invention was sufficiently original that some historians have called him one of the most important inventors in all of human history.

James Watt was born on January 19, 1763, at Greenock, Renfrewshire, Scotland, on the Firth of Clyde. His mother, Agnes Muirhead Watt, came from a distinguished family that dated its origin to the early 12th century, and his father, was a shipwright and contractor, who also served as the town's chief baillie, a position roughly comparable to that of a magistrate. Young James was educated largely at home, although he eventually attended the Greenock School in his later childhood years. At the age of 18, he left home to learn instrument-making in London. A year later, he returned to Scotland, intending to establish his own instrument-making and repair business. In a stroke of good fortune, the University of Glasgow asked him to repair and restore a group of astronomical instruments that had come into its possession. His work was so satisfactory, that he was taken on as an employee at the university and provided with a small workshop. One of the professors with whom he worked at the university was the eminent chemist Joseph Black (1728–1799), who had himself conducted important research on the latent heat of materials, an important theoretical concept involved in the operation of steam engines.

Watt's introduction to the steam engine came in 1763 when he was asked to repair a Newcomen engine owned by the university. He had heard about the steam engine four years earlier, but had been unsuccessful in building such a device. During his work on the Newcomen engine, however, Watt recognized the fundamental problem inherent in the Newcomen design, the heating–cooling–reheating cycle that diminished the machine's efficiency. He eventually designed and built his own version of the steam engine, one that rapidly became very popular, especially for use in the pumping of water from mines.

Watt's efforts to commercialize his invention did not go smoothly. At first, he made an arrangement with one John Roebuck, a well-known and successful inventor and industrialist. The two encountered problems almost immediately, however, in finding craftsmen who could provide the components needed to mass produce the invention. After struggling for eight years (during which time Watt worked as a civil engineer and surveyor), Roebuck's own business went bankrupt, and, in 1773, he sold his shares in the business to Matthew Boulton, who owned a foundry that was able to supply the skilled workers needed to make Watt's invention a reality and a commercial success. With this change, production of the Watt's steam engine began in earnest, and orders were soon pouring in for the new device. Over the next quarter century, Watt continued to make a number of modifications and improvements on the steam engine, and by the time he retired from the business in 1800, more than 500 steam engines were operating in mines and factories throughout the British Isles.

By the time he retired, Watt had become a rich man, and he spent the rest of his life traveling with his second wife and inventing. Most of his inventions were clever, but not particularly useful. They included a steam mangle, a device for measuring distances using a telescope, and an improvement in the common oil lamp. Watt died on August 25, 1819, at his home, "Heathfield," in Handsworth, Birmingham, England. He was buried next to his long-time partner, Matthew Boulton, at St. Mary's Church in Handsworth. Watt's contributions to science and society have been recognized in a number of ways. In 1908, the new Watt Memorial Engineering and Navigation School (now the James Watt Memorial College) was named in his honor. The unit of power in the International System of Units (SI) was also named in his honor. A watt is a joule of energy per second of time. On November 2, 2011, the Bank of England introduced a new 50-pound note featuring a drawing of Watt and Boulton, the first time a bank note has contained the feature of two British subjects.

Wave Energy Centre (Centro de Energia das Ondas)
Av. Manuel da Maia, 36, r/c Dto.
1000-201 Lisboa, Portugal
Phone: +351 21 848 2655
Fax: +351 21 848 1630
E-mail: mail@wave-energy-centre.org
http://www.wavec.org/index.php/1/home/

The Wave Energy Centre (WavEC) had a modest beginning in 2001 when an organization designed to study the potential applications of ocean and wave energy was proposed at an international conference, "Oceans III Millenium," the first International Congress on Marine Science and Technology, held in Pontevedra, Spain. Two years later, the organization came into existence with a staff of four, a director, a secretary, and two researchers from the Faculty of Technology at the Instituto Superior Técnico, along with 10 associates. Today the organization has grown to include 19 specialists in fields such as hydrodynamics, numerical modeling, environmental assessment, monitoring, and communication. WavEC works closely with a number of associate organizations interested in a variety of ocean-wave related research, including A. Silva Matos Energia, producers of equipment used in alternative energy facilities; Consulmar—Projectistas e Consultores Limitada, a planning and consulting firm with a number of coastal and maritime projects in Portugal and other locations; EDA—Electricidade dos Açores, the electrical utility for the Azores Islands; Generg Sociedade Gestora de Participações Sociais, SA, a company that designs and constructs alternative energy plants; and Kymaner, Tecnologias Energéticas Lda, a company founded to develop and promote wave energy in Portugal.

WavEC's work falls into five major areas: monitoring, technologies, numerical modeling, environment, and training. Monitoring programs aim to evaluate new technologies as they are developed for implementation. Technologies focus on the analysis and evaluation of various methods for taking advantage of ocean wave energy. Numerical modeling provides theoretical tools for assessing the potential value of various technologies. Environment involves a study of potential environmental impacts of various types of wave energy technologies. And training is a program for providing students with the technical skills to conduct research and work within the wave energy field.

Since 2000, WavEC has been involved in about 15 major projects related to the development of wave energy technology. As an example, Aqua-RET (Aquatic Renewable Energy Technologies) was funded by the European Commission for 2006–2008 for the purpose of developing online learning opportunities for the general public and companies on the potential of wave energy as a form of alternative energy. MaRINET (Marine Renewables Infrastructure Network for Emerging Energy) is a large-scale project that will run from 2011 to 2015 bringing together information about a variety of wave energy project, ranging from small models to large prototypes. SOWFIA (Streamlining of Ocean Wave Farms Impact Assessment) was in operation from 2010 to 2012. It studied a variety of wave energy farms in a number of cooperating European nations.

In addition to these projects, WavEC sponsors and takes part in a number of meetings and other events annually, such as the THETIS MRE: International Convention in Marine Renewable Energy; 4th CoastLab Teaching School—Wave and Tidal Energy (short course on ocean energy); Conference on Marine Economy—A key component of the EU Integrated Strategy for the Atlantic Area; Wavetrain2 final conference and course on Socio-Economic Impacts of Offshore Renewables; and Oceanology International 2012.

In addition to its own web page, the Wave Energy Centre maintains other web pages that provide more detailed information and photographs on specific wave energy projects, such as the 400 kilowatt OWC Pico Power Plant located on Pico Island in the Azores (http://www.pico-owc.net/). WavEC's website also has a valuable library with reports, papers, and presentations on wave energy topics as well as videos of their work and links to many other organizations interested in the issue of ocean and wave energy.

World Coal Association

5th Floor, Heddon House
149-151 Regent Street
London
W1B 4JD
United Kingdom
Phone: +44 20 7851 0052
Fax: +44 20 7851 0061
E-mail: info@worldcoal.org
http://www.worldcoal.org/

The World Coal Association (WCA) is a nonprofit, nongovernmental organization the purpose of which is to represent the interests of the coal industry in discussions about the resource around the world. The WCA was founded in 1985 as the International Coal Development Institute, and changed its name to the World Coal Institute in 1988, and then to its present name in 2010. The organization states that the coal industry is in need of a united front that can deal with the challenges facing this energy source in coming decades. It has expressed a special concern about reversing public images of coal primarily as a source of carbon dioxide emissions and more as an essential element in the solution of the world's energy needs over the next century or more. The two main elements of this philosophy are that:

- "coal as a strategic resource that is widely recognised as essential for a modern quality of life, a key contributor to sustainable development, and an essential element in enhanced energy security; and
- the coal industry as a progressive industry that is recognised as committed to technological innovation and improved environmental outcomes within the context of a balanced and responsible energy mix."

In order to achieve its objectives, the WCA works with governmental agencies and organizations, scientific and technical associations, coal producers and consumers, nongovernmental groups, the public media, environmental organizations, and any other associations interested in coal-related issues. In addition to fundamental questions about the economics of coal production and consumption, WCA deals with problems of global climate change, sustainable development, and local issues related to coal mining and coal use.

Two categories of membership are available in WCA, corporate members and associate members. Corporate members are organizations involved in the production, sale, transport, and use of coal, along with suppliers of equipment in these fields. Some current corporate members are Anglo Coal; Arch Coal, Incorporated; PHPbilliton; Coal India; Consol Energy; Glencore; Joy Mining Machinery; Mitsubishi Development, Peabody Energy; Rio Tinto; and Solid Energy Coals of New Zealand. Associate members are nonprofit organizations with a special interest in the coal industry, coal equipment industry, power sector, and research arm of energy organizations. Some current associate members are the Association of UK Coal Importers; the Coal Association of Canada; Australian Coal Association; Confederation of UK Coal Producers; Coal Association of New Zealand; China National Coal Association; Global CCS Institute; and VGB Power Tech.

A major activity of the WCA is to form working alliances with a number of energy-related national and international institutions, such as the UN Commission on Sustainable Development, the UN Framework Convention on Climate Change; the International Energy Agency Working Group of Fossil Fuels; the International Energy Agency Working Group Coal Industry Advisory Board; and the Carbon Sequestration Leadership Forum. The organization also organizes workshops that bring together representatives from government, industry, and interested nongovernmental organizations. WCA is also actively involved in lobbying governmental agencies on policy stands associated with the production and consumption of coal and coal products. WCA publishes a quarterly newsletter, *ECOAL,* which is available only electronically. The association has also produced a number of reports, brochures, and other publications, such as CCS & the Clean Development Mechanism; Case Studies; Coal: Delivering Sustainable Development; Coal: Liquid Fuels; Coal Meeting the Climate Challenge—Technology to Reduce GHG Emissions; Coal Meeting Global Challenges; The Coal Resource—A Comprehensive Overview of Coal; Coal: Secure Energy; Coal and Steel; and Key Elements of a Post-2012 Agreement on Climate Change. Some of these publications are available in hard cover, while most are available only electronically through the association's website at http://www.worldcoal.org/resources/wca-publications/.

The WCA website is an especially rich source of information on all aspects of the use of coal for the production of energy. It contains an excellent general introduction to the science and technology of coal, including sections on the properties of coal, production and marketing of coal, methane production from coal seams, and where coal is found. Another section deals with carbon capture and storage, the technology by which carbon dioxide produced during combustion is prevented from escaping into the atmosphere. The technology is one of the most promising ways of significantly reducing the release of carbon dioxide to the atmosphere as a result of burning coal and other fossil fuels. A third section discusses environmental aspects of coal production and use, including coal mining and pollution produced by the use of coal as a fuel. A final section focuses on coal and society, with a review of the role that coal plays in various societies around the world, as well as safety issues in the production of coal and coal and energy security issues. The website also contains archived news articles, a coal-focused blog, frequently asked questions about coal production and consumption, and links to other resources with information about coal-related issues.

World Energy Council
5th Floor, Regency House
1-4 Warwick Street
London W1B 5LT
United Kingdom
Phone: +44 20 7734 5996
Fax: +44 20 7734 5926
E-mail: info@worldenergy.org
http://www.worldenergy.org/

The World Energy Council was established in 1923, largely through the efforts of Daniel N. Dunlop, a Scotsman somewhat peripherally involved with the electrical business (he was, for some years, manager of the Westinghouse Electrical Company's European Publicity Department). Dunlop felt that an international association was needed to assess national and worldwide energy issues and to consider policies that would advance the development of energy use in future years. In response to Dunlop's call, the First World Power Conference convened in London in 1924, a conference attended by 1,700 delegates from 40 countries. There was sufficient interest expressed at the conference to lead to the formation of a new organization, also called the World Power Conference, on July 11, 1924. That organization consisted primarily of a number of National Member Committees, whose activities were coordinated by an International Executive Council, chaired by Dunlop. The program of the conference was set out in a set of "Objects," which were later revised in 1958 and 1968. In the latter year, at the association's annual meeting in Moscow, its name was also changed to the World Energy Conference (WEC). It adopted its present name at the 1992 conference held in Madrid. Much of the day-to-day operations of the association throughout most of its history were under the supervision of the Secretary of the conference, a post held for 40 years by Charles Gray. That position within the organization is now called the Secretary General, a post currently held by Swiss electrical engineer Christoph Frei.

In 1978, the WEC decided to abandon its historic program of annual meetings to be replaced by triennial congresses, a pattern it still follows. At each triennial congress, the organization adopts a three-year business plan designed to guide its activities between congresses. The 2011–2013 program consists of six primary components, known as Activity Areas. Three of these Activity Areas deal with long-term (strategic) issues: Energy Resources and Technology, Energy and Climate National Policies, and Global Energy Scenarios. Three other Activity Areas deal with short-term issues: Energy Access, Energy and Urban Innovation, and Global Energy Frameworks. WEC calls on a number of Knowledge Networks that provide basic information that can be used across specific Activity Areas. These Knowledge Networks include topics such as cleaner fossil fuel systems, energy and mobility, energy efficiency, innovative financing mechanisms, performance of power generating plants, and rules of trade.

The Activity Areas selected for the 2011–2013 program largely defined the programmatic efforts of the WEC during the defined period. For example, the long-term Activity Area called Energy Resources and Technologies includes topics such as when the world will run out of oil; the current status of shale gas, biomass, wind, solar, and other renewable energy resources, as well as other fossil fuel sources; issues related to the use of smart grids, energy-water linkages, carbon capture and sequestration, and clean coal and nuclear technologies; and the potential balance between energy return compared to energy investment. The short-term Activity Area called Energy Access focuses on problems such as the development of financing systems that will make possible the provision of adequate energy supplies in rural areas; improved education for the general public about energy issues; improved understanding of and political support for wider local control over energy resources; and development of skills within local communities that allow them to deal with their unique energy problems.

WEC currently has a membership of 91 nations. The most prominent nonmember is Australia, although some South American, African, and Southeast Asian nations are also nonmembers. General policy decisions are made at the triennial congress, the most recent of which was held in Montréal in 2010. The next congress is scheduled to be held in Daegu, South Korea, on October 13–17, 2013. Between triennial congresses, annual conferences of the Executive Assembly are held to make short-term policy and decisions. The Executive Assembly consists of cabinet-level representatives from all member states, and it is the ultimate authority on the governance of council activities. The council's articles of association authorize the Executive Assembly to carry out a number of acts, such as issuing general policy statements, approving applications for membership, approving the organization's budget and business plan, approving the plan and budget for the council's work program, and establishing committees to carry out the organization's work program.

In addition to the triennial congress and annual meetings of the Executive Assembly, WEC sponsors and participates in a number of local and regional meetings on a variety of energy-related topics. For example, in 2011, WEC-related meetings included participation in the European Union Sustainable Energy Week, North Sea Offshore Grid CEO Discussions, Carbon Sequestration Leadership Forum Ministerial Meeting, Portuguese conference on nonconventional oil and gas; the WEC Colombian Member Committee Event; and the WEC Italy Conference.

The council annually publishes a large number of reports on energy issues. Some examples include *Survey of Energy Resources 2010; World Energy Insight 2010; Roadmap towards a Competitive European Energy Market* (2010)*; Logistics Bottlenecks* (2010)*; Energy Efficiency: A Recipe for Success* (2010)*; Water for Energy* (2010)*; Shale Gas* (2010)*; Energy and Urban Innovation* (2010)*; Pursuing Sustainability: 2010 Assessment of Country Energy and Climate Policies; Biofuels: Policies, Standards and Technologies 2010; European Climate Change Policy Beyond 2012; Energy Policy Scenarios to 2050;* and *Trade and Investment Rules for Energy.*

Daniel Yergin (1947–)

Daniel Yergin is one of the best-known writers about energy-related issues in the world today. His 1991 book, *The Prize: The Epic Quest for Oil, Money, and Power,* won the 1992 Pulitzer Prize for General Non-Fiction. The book reached number 1 on the *New York Times* best seller list on February 10, 1991, and has been translated into 17 languages. *Business Week* magazine called *The Prize* "the best history of oil ever written." The book was eventually made into a PBS miniseries that was reputedly seen by more than 20 million viewers. Yergin followed up on *The Prize* in 2011 with a second book on a world of the future that was no longer dependent on the petroleum industry called *The Quest: Energy, Security, and the Remaking of the Modern World,* which a *New York Times* reviewer called "if anything . . . an even better book . . . [that] will be necessary reading for C.E.O.'s, conservationists, lawmakers, generals, spies, tech geeks, thriller writers, ambitious terrorists and many others."

Daniel Howard Yergin was born on February 6, 1947, in Los Angeles, California. His parents were Irving H. Yergin, a reporter for the *Chicago Tribune,* and Naomi Yergin, a painter and a sculptor. He graduated from Beverly Hills High School, where he was student body president, in 1964 and then attended Yale University, from which he received his BA degree in English in 1968. While at Yale, he served on the board of the *Yale Daily News.* After graduation, Yergin won a Marshall Scholarship that allowed him to continue his studies at England's University of Cambridge, where he earned first a second bachelor's degree, this time in history, and then a PhD in international relations in 1974. During his years at Cambridge, Yergin also served as contributing editor for the *New York* magazine.

Upon graduation from Cambridge, Yergin took a position as a research fellow at Harvard University from 1974 to 1976, followed by a similar position at the Rockefeller Foundation from 1975 to 1979. Concurrently, he was also lecturer at the Harvard Business School from 1976 to 1979. In 1979, Yergin moved to the John F. Kennedy School of Government at Harvard, where he served as a lecturer until 1983. During the first years of that appointment, he also worked for the Solar Energy Research Institute in Golden, Colorado. In 1983, Yergin cofounded (with James Rosenfeld) and became president of the Cambridge Energy Research Associates (CERA), a firm that offers consulting services to governmental agencies and private industries on a variety of energy-related issues. CERA was acquired by IHS, Inc., of Douglas County, Colorado, a private information services corporation in 2004. Yergin continues to serve as president of IHS CERA as of early 2012.

Yergin has been a prolific speaker and writer about energy and political issues over the past three decades. His first book, *Shattered Peace: The Origins of the Cold War and the National Security State* (1977) was based on his doctoral thesis. It was followed by *Energy Future: The Report of the Energy Project at the Harvard Business School* (1979, with Robert B. Stobaugh), *The Dependence Dilemma: Gasoline Consumption and America's Security* (1980), *Global Insecurity: A Strategy for Energy and Economic Renewal* (1982, with Martin Hillenbrand), *The U.S. Strategic Petroleum Reserve* (1990), *Gasoline and the American People* (1991), *Russia 2010: And What It Means for the World* (1993, with Thane Gustafson), *The Commanding Heights: The Battle Between Government and the Marketplace That Is Remaking the Modern World* (1998, with Joseph A. Stanislaw), *The Euro: Remaking Europe's Future* (1998), and *Global Energy: Challenges and Priorities* (2001, with Amory Lovins and Dennis Eklof).

Yergin has also served in a number of professional positions, including the board of directors of the Marshall Scholars, the New America Foundation, and the U.S. Energy Associates; as a fellow of the German Marshall Fund and the Atlantic Institute for International Affairs; on the advisory panel of U.S.–Japan Relations at Harvard; and a member of the American History Association, the Council on Foreign Relations, the National Petroleum Council, and in the International Panel of Advisors of the Asia-Pacific Petroleum Conference. He also served as chair of the U.S. Department of Energy's Task Force on Strategic Energy Research and Development, is a trustee of the Brookings Institution, on serves on the advisory board of the Massachusetts Institute of Technology Energy Initiative. In 1997 he received the U.S. Energy Award for "lifelong achievements in energy and the promotion of international understanding." In 1994 he was awarded an honorary doctoral degree from the University of Houston.

5 Data and Documents

This chapter provides some relevant data and documents dealing with the worldwide energy crisis. An important source for much of this information is reports that have been conducted by national governments, energy companies, and interested nongovernmental organizations.

Data

The following tables summarize the production and consumption of oil in a number of countries throughout the world between 1999 and 2009.

An oil tanker is moored at an oil loading platform adjacent to an oil refinery in Kawasaki, west of Tokyo. (AP Photo/Koji Sasahara, File)

Table 5.1 Production of Petroleum (in Thousands of Barrels per Day)*

	1999	2000	2001	2002	2003	2004	2005	2006	2007	2008	2009
North America											
U.S.	7731	7733	7669	7626	7400	7228	6895	6841	6847	6734	7196
Canada	2604	2721	2677	2858	3004	3085	3041	3208	3320	3268	3212
Mexico	3343	3450	3560	3585	3789	3824	3760	3683	3471	3167	2979
South and Central America											
Argentina	847	819	830	818	806	754	725	716	699	682	676
Brazil	1133	1268	1337	1499	1555	1542	1716	1809	1833	1899	2029
Colombia	838	711	627	601	564	551	554	559	561	616	685
Ecuador	383	409	416	401	427	535	541	545	520	514	495
Peru	107	100	98	98	92	94	111	116	114	120	145
Trinidad & Tobago	141	138	135	155	164	152	171	174	154	149	151
Venezuela	3126	3239	3142	2895	2554	2907	2937	2808	2613	2558	2437
Others	124	130	137	152	153	144	143	141	143	140	141
Europe and Eurasia											
Azerbaijan	279	282	301	311	313	315	452	654	869	915	1033
Denmark	299	363	348	371	368	390	377	342	311	287	265
Italy	104	95	86	115	116	113	127	120	122	108	95
Kazakhstan	631	744	836	1018	1111	1297	1356	1426	1484	1554	1682
Norway	3139	3346	3418	3333	3264	3189	2969	2779	2550	2451	2342

Romania	133	131	130	127	123	119	114	105	99	98	93
Russia	6178	6536	7056	7698	8544	9287	9552	9769	9978	9888	10032
Turkmenistan	143	144	162	182	202	193	192	186	198	205	206
United Kingdom	2909	2667	2476	2463	2257	2028	1809	1636	1638	1526	1448
Uzbekistan	191	177	171	171	166	152	126	125	114	114	107
Others	474	465	465	501	509	496	468	455	448	425	400
Middle East											
Iran	3603	3855	3892	3709	4183	4248	4234	4286	4322	4327	4216
Iraq	2610	2614	2523	2116	1344	2030	1833	1999	2143	2423	2482
Kuwait	2085	2206	2148	1995	2329	2475	2618	2690	2636	2782	2481
Oman	911	959	960	904	824	786	778	742	715	754	810
Qatar	723	757	754	764	879	992	1028	1110	1197	1378	1345
Saudi Arabia	8853	9491	9209	8928	10164	10638	11114	10853	10449	10846	9713
Syria	579	548	581	548	527	495	450	435	415	398	376
United Arab Emirates	2511	2547	2455	2260	2553	2664	2753	2971	2900	2936	2599
Yemen	405	450	455	457	448	420	416	380	345	304	298
Other	48	48	47	48	48	48	34	32	35	33	37

(Continued)

Table 5.1 *(continued)*

	1999	2000	2001	2002	2003	2004	2005	2006	2007	2008	2009
Africa											
Algeria	1515	1578	1562	1680	1852	1946	2015	2003	2016	1993	1811
Angola	745	746	742	905	870	1103	1405	1421	1684	1875	1784
Cameroon	95	88	81	72	67	89	82	87	82	84	73
Chad	–	–	–	–	24	168	173	153	144	127	118
Republic of Congo (Brazzaville)	266	254	234	231	215	216	246	262	222	249	274
Egypt	827	781	758	751	749	721	696	697	710	722	742
Gabon	340	327	301	295	240	235	234	235	230	235	229
Libya	1425	1475	1427	1375	1485	1623	1745	1815	1820	1820	1652
Nigeria	2066	2155	2274	2103	2238	2431	2499	2420	2305	2116	2061
Sudan	63	174	217	241	265	301	305	331	468	480	490
Tunisia	84	78	71	74	68	71	73	70	97	89	86
Other	56	56	53	63	71	75	72	66	84	79	79
Asia/Pacific											
Australia	625	809	733	730	624	582	580	554	567	556	559
Brunei	182	193	203	210	214	210	206	221	194	175	168
China	3213	3252	3306	3346	3401	3481	3627	3684	3743	3901	3790
India	736	726	727	753	756	773	738	762	769	768	754
Indonesia	1408	1456	1389	1289	1183	1129	1087	1017	969	1031	1021
Malaysia	737	735	719	757	776	793	759	747	763	768	740

Thailand	140	176	191	204	236	223	265	286	305	321	330
Vietnam	296	328	350	354	364	427	398	367	337	317	345
Other	218	200	195	193	195	235	286	305	320	340	328
Total											
World	72325	74820	74813	74533	76916	80371	81261	81557	81446	81995	79948
of which:											
European Union	3684	3493	3285	3339	3128	2902	2659	2422	2388	2222	2082
OECD	21103	21521	21303	21430	21165	20766	19861	19458	19140	18414	18390
OPEC	29646	31072	30544	29132	30877	33592	34721	34920	34604	35568	33076
Non-OPEC	35127	35734	35608	35869	35540	35371	34700	34321	34046	33602	33671
Former Soviet Union	7552	8014	8660	9533	10499	11407	11839	12316	12795	12825	13202

Consumption (thousands of barrels per day)**

	1999	**2000**	**2001**	**2002**	**2003**	**2004**	**2005**	**2006**	**2007**	**2008**	**2009**
North America											
U.S.	19519	19701	19649	19761	20033	20732	20802	20687	20680	19498	18686
Canada	1926	1937	2023	2067	2132	2248	2247	2246	2323	2287	2195
Mexico	1842	1910	1899	1837	1885	1918	1974	1970	2017	2010	1945
South & Central America											
Argentina	445	431	405	364	372	388	414	431	484	499	473
Brazil	2114	2056	2082	2063	1985	1999	2033	2087	2258	2397	2405

(Continued)

Table 5.1 (continued)

	1999	2000	2001	2002	2003	2004	2005	2006	2007	2008	2009
Chile	249	236	230	228	228	240	254	264	346	357	333
Colombia	237	233	216	211	211	214	229	238	229	199	194
Ecuador	131	129	132	131	137	141	168	182	196	207	216
Peru	159	155	148	147	140	153	152	147	154	172	188
Venezuela	474	496	545	594	479	523	576	607	597	607	609
Other	1095	1118	1159	1175	1201	1213	1222	1254	1269	1243	1235
Europe and Eurasia											
Austria	251	245	265	272	294	286	295	294	278	279	270
Azerbaijan	111	123	81	74	86	92	108	98	92	74	60
Belarus	154	143	149	145	148	153	146	165	152	172	192
Belgium & Luxembourg	670	702	669	691	748	785	815	839	832	812	781
Bulgaria	93	84	87	98	115	105	109	116	113	103	98
Czech Republic	174	169	179	174	185	203	211	208	206	210	205
Denmark	222	215	205	200	193	189	195	197	196	189	174
Finland	224	224	222	226	239	224	233	225	226	225	212
France	2044	2007	2023	1967	1965	1978	1960	1956	1923	1902	1833
Germany	2824	2763	2804	2714	2664	2634	2605	2624	2393	2517	2422
Greece	384	407	412	416	405	438	436	454	445	437	417
Hungary	151	145	142	140	138	142	163	169	169	164	161
Iceland	18	19	18	19	18	20	21	20	24	20	20

Ireland	172	170	185	182	178	184	194	194	198	190	169
Italy	1980	1956	1946	1943	1927	1873	1819	1813	1759	1680	1580
Kazakhstan	147	158	180	195	207	227	234	239	245	263	260
Lithuania	63	49	56	53	51	55	58	59	59	64	61
Netherlands	880	897	942	952	962	1003	1070	1093	1144	1089	1054
Norway	216	201	213	208	219	210	212	217	222	214	211
Poland	431	427	415	420	435	460	479	516	535	554	553
Portugal	330	324	327	338	317	322	331	300	302	283	269
Romania	195	203	217	226	199	230	223	219	223	221	211
Russia	2625	2583	2566	2606	2622	2619	2601	2709	2708	2817	2695
Slovakia	73	73	68	76	71	68	81	82	86	90	83
Spain	1423	1452	1508	1526	1559	1593	1619	1602	1617	1574	1492
Sweden	337	318	318	317	332	319	315	322	308	302	287
Switzerland	271	263	281	267	259	258	262	269	243	258	262
Turkey	638	677	645	656	662	667	656	635	656	663	621
Turkmenistan	80	79	83	86	95	95	100	118	113	117	120
Ukraine	272	256	288	286	295	310	296	309	339	336	307
UnitedKingdom	1721	1697	1697t	1693	1717	1764	1802	1785	1714	1681	1611
Uzbekistan	138	132	130	125	142	134	109	102	99	101	101
Other	448	417	445	469	493	502	540	549	582	592	580

(Continued)

Table 5.1 *(continued)*

	1999	2000	2001	2002	2003	2004	2005	2006	2007	2008	2009
Middle East											
Iran	1221	1301	1314	1413	1498	1558	1620	1693	1685	1761	1741
Kuwait	243	246	251	271	296	327	359	333	338	370	419
Qatar	51	60	73	89	105	122	144	158	174	198	209
SaudiArabia	1543	1579	1605	1632	1759	1880	1987	2065	2212	2390	2614
United Arab Emirates	271	258	297	326	340	364	394	420	448	475	455
Other	1358	1394	1439	1432	1396	1454	1506	1578	1612	1671	1708
Africa											
Algeria	187	192	200	222	231	240	251	260	288	311	331
Egypt	573	564	548	534	550	567	629	610	650	693	720
South Africa	457	463	474	486	503	514	516	529	550	532	518
Other	1273	1264	1295	1310	1329	1369	1405	1387	1442	1509	1513
Asia/Pacific											
Australia	843	837	845	846	851	856	886	918	925	936	941
Bangladesh	68	66	80	80	83	83	94	93	93	92	93
China	4477	4772	4872	5288	5803	6772	6984	7410	7771	8086	8625
Hong Kong	194	202	244	268	270	316	287	305	324	294	286
India	2134	2254	2284	2374	2420	2573	2569	2580	2838	3071	3183
Indonesia	1019	1122	1162	1191	1218	1290	1289	1252	1273	1314	1344
Japan	5598	5557	5422	5347	5440	5269	5343	5213	5039	4846	4396

Malaysia	435	435	442	482	473	485	469	458	481	476	468
New Zealand	130	133	134	139	149	149	154	156	156	156	148
Pakistan	363	373	366	357	321	325	312	356	388	389	
Philippines	375	348	347	331	331	338	315	284	300	265	265
Singapore	619	654	716	699	668	748	794	853	916	968	1002
South Korea	2178	2229	2235	2282	2300	2283	2308	2317	2389	2287	2327
Taiwan	964	1003	991	999	1069	1084	1090	1097	1123	1037	1014
Thailand	788	784	768	827	881	967	1005	996	985	962	975
Other	334	357	375	381	396	419	432	433	462	481	516
Total											
World	75648	76428	77032	77945	79424	82261	83513	84367	85619	85239	84077
of which:											
European Union	14814	14692	14861	14797	14868	15032	15204	15260	14926	14775	14143
OECD	47469	47653	47692	47676	48277	49073	49489	49323	49008	47353	45327
Former Soviet Union	3714	3631	3646	3688	3769	3815	3798	3948	3973	4115	3965
Other Emerging Market Economies	24465	25144	25694	26581	27377	29374	30226	31096	32639	33771	34785

* Includes crude oil, shale oil, oil sands and NGLs (the liquid content of natural gas where this is recovered separately). Excludes liquid fuels from other sources such as biomass and coal derivatives.

** Inland demand plus international aviation and marine bunkers and refinery fuel and loss. Consumption of fuel ethanol and biodiesel is also included.

Source: BP Statistical Review of World Energy 2010, BP p.l.c. Available online at http://www.bp.com/liveassets/bp_internet/globalbp/globalbp_uk_english/reports_and_publications/statistical_energy_review_2008/STAGING/local_assets/2010_downloads/statistical_review_of_world_energy_full_report_2010.pdf. Accessed March 7, 2012. Used by permission.

The following table summarizes the extent of coal reserves in major coal-producing nations around the world.

Table 5.2 Proved Reserves of Coal at the End of 2009 (in Millions of Tonnes)

Country	Anthracite and Bituminous	Sub-bituminous and Lignite	Total
United States	108,950	129,358	238,308
Canada	3471	3107	6578
Mexico	860	351	1211
Total North America	113,281	132,816	246,097
Brazil	–	7059	7059
Colombia	6434	380	6814
Venezuela	479	–	479
Other South and Central America	51	603	654
Total South and Central America	6964	8042	15,006
Bulgaria	5	1991	1996
Czech Republic	1673	2828	4501
Germany	152	6556	6708
Greece	–	3900	3900
Hungary	199	3103	3302
Kazakhstan	28,170	3130	31,300
Poland	6012	1490	7502
Romania	12	410	422
Russian Federation	49,088	107,922	157,010
Spain	200	330	530
Turkey	–	1814	1814
Ukraine	15,351	18,522	33,873
United Kingdom	155	–	155
Other Europe and Eurasia	1025	18,208	19,233
Total Europe and Eurasia	102,042	170,204	272,246
South Africa	30,408	–	30,408
Zimbabwe	502	–	502

(continued)

Table 5.2 (*continued*)

Country	Anthracite and Bituminous	Sub-bituminous and Lignite	Total
Other Africa	929	174	1103
Middle East	1386	–	1386
Total Middle East and Africa	33,225	174	33,399
Australia	36,800	39,400	76,200
China	62,200	52,300	114,500
India	54,000	4600	58,600
Indonesia	1721	2607	4328
Japan	355	–	355
New Zealand	33	538	571
North Korea	300	300	600
Pakistan	1	2069	2070
South Korea	133	–	133
Thailand	–	1354	1354
Vietnam	150	–	150
Other Asia Pacific	115	276	391
Total Asia Pacific	155,809	103,444	259,253
Total World	411,321	414,680	826,001
of which:			
European Union	8427	21143	29570
OECD	159,012	193,083	352,095
Former Soviet Union	93,609	132,386	225,995
Other EMEs	158,700	89,211	247,911

Note: Proved reserves of coal. Generally taken to be those quantities that geological and engineering information indicates with reasonable certainty can be recovered in the future from known deposits under existing economic and operating conditions.

Source: BP Statistical Review of World Energy 2010, BP p.l.c. Available online at http://www.bp.com/liveassets/bp_internet/globalbp/globalbp_uk_english/reports_and_publications/statistical_energy_review_2011/STAGING/local_assets/spreadsheets/statistical_review_of_world_energy_full_report_2011.xls#'Coal Reserves'!A1. Accessed March 7, 2012. Used by permission.

The following table summarizes proved natural gas reserves in a number of countries around the world.

Table 5.3 Proved Reserves of Natural Gas (in Trillion Cubic Meters)

Country	Reserves	Percent of World Total
United States	6.93	3.7
Canada	1.75	0.9
Mexico	0.48	0.3
Total North America	9.16	4.9
Argentina	0.37	0.2
Bolivia	0.71	0.4
Brazil	0.36	0.2
Colombia	0.12	0.1
Peru	0.32	0.2
Trinidad & Tobago	0.44	0.2
Venezuela	5.67	3.0
Other South and Central America	0.07	–
Total South and Central America	8.06	4.3
Azerbaijan	1.31	0.7
Denmark	0.06	–
Germany	0.08	–
Italy	0.06	–
Kazakhstan	1.82	1.0
Netherlands	1.09	0.6
Norway	2.05	1.1
Poland	0.11	0.1
Romania	0.63	0.3
Russian Federation	44.38	23.7
Turkmenistan	8.10	4.3
Ukraine	0.98	0.5
United Kingdom	0.29	0.2
Uzbekistan	1.68	0.9
Other Europe and Eurasia	0.44	0.2
Total Europe and Eurasia	63.09	33.7
Bahrain	0.09	–
Iran	29.61	15.8
Iraq	3.17	1.7
Kuwait	1.78	1.0
Oman	0.98	0.5
Qatar	25.37	13.5
Saudi Arabia	7.92	4.2
Syria	0.28	0.2
United Arab Emirates	6.43	3.4

(continued)

Table 5.3 *(continued)*

Country	Reserves	Percent of World Total
Yemen	0.49	0.3
Other Middle East	0.06	—
Total Middle East	76.18	40.6
Algeria	4.50	2.4
Egypt	2.19	1.2
Libya	1.54	0.8
Nigeria	5.25	2.8
Other Africa	1.27	0.7
Total Africa	14.76	7.9
Australia	3.08	1.6
Bangladesh	0.36	0.2
Brunei	0.35	0.2
China	2.46	1.3
India	1.12	0.6
Indonesia	3.18	1.7
Malaysia	2.38	1.3
Myanmar	0.57	0.3
Pakistan	0.91	0.5
Papua New Guinea	0.44	0.2
Thailand	0.36	0.2
Vietnam	0.68	0.4
Other Asia Pacific	0.36	0.2
Total Asia Pacific	16.24	8.7
Total World	187.49	100.0
of which:		
European Union	2.42	1.3
OECD	16.18	8.6
Former Soviet Union	58.53	31.2

Note: Proved reserves of natural gas. Generally taken to be those quantities that geological and engineering information indicates with reasonable certainty can be recovered in the future from known reservoirs under existing economic and operating conditions.

Source: BP Statistical Review of World Energy 2010, BP p.l.c. Available online at http://www.bp.com/assets/bp_internet/globalbp/globalbp_uk_english/reports_and_publications/statistical_energy_review_2011/STAGING/local_assets/spreadsheets/statistical_review_of_world_energy_full_report_2011.xls#'Gas—Proved reserves'!A1. Accessed March 7, 2012. Used by permission.

The following tables show the 2011 projections by the U.S. Energy Information Administration for renewable energy supplies from 2015 to 2035.

Table 5.4 Projected Renewables, 2015–2035[1]

Hydroelectric							
Region/Country[2]	**2008**	**2015**	**2020**	**2025**	**2030**	**2035**	**% Change**
United States	78	78	78	79	80	81	0.1
Canada	74	78	81	86	95	100	1.1
Mexico/Chile	16	19	21	23	27	31	2.4
OECD Europe	150	157	170	178	180	180	0.7
Japan	22	24	24	24	24	24	0.4
South Korea	2	2	2	2	2	2	0.1
Australia/New Zealand	13	13	13	13	14	14	0.3
Russia	47	53	57	61	66	72	1.6
China	172	247	318	327	335	360	2.8
India	39	56	80	85	96	106	3.8
Brazil	78	89	104	127	154	173	3.0
Other Eurasia	40	42	46	46	47	50	0.8
Other Asia	39	61	81	89	98	114	4.1
Other Central/South America	53	63	71	78	85	91	2.0
Middle East	12	14	17	18	19	21	2.0
Africa	22	28	32	35	40	45	2.7
WORLD	857	1,025	1,195	1,272	1,362	1,463	2.0
Wind Power							
Region/Country[2]	**2008**	**2015**	**2020**	**2025**	**2030**	**2035**	**% Change**
United States	25	51	51	54	55	57	3.1
Canada	2	11	13	14	15	17	7.5

Mexico/Chile	0	5	5	5	6	7	16.81
OECD Europe	65	126	181	207	217	227	4.8
Japan	2	4	5	8	8	8	5.6
South Korea	0	1	2	2	3	4	9.9
Australia/New Zealand	2	8	11	11	11	11	6.2
Russia	0	0	0	0	0	0	0.1
China	12	62	99	119	139	156	9.9
India	10	14	16	20	22	24	3.3
Brazil	0	2	2	3	3	4	8.5
Other Eurasia	0	4	4	4	5	5	9.8
Other Asia	0	1	3	3	4	4	10.51
Other Central/South America	0	1	1	1	1	1	5.0
Middle East	0	1	1	1	1	2	11.31
Africa	0	4	5	5	6	6	10.3
WORLD	121	293	398	456	496	533	5.7

Geothermal Power

Region/Country[2]	**2008**	**2015**	**2020**	**2025**	**2030**	**2035**	**% Change**
United States	2	3	3	4	6	6	3.7
Canada	0	0	0	0	0	0	0.0

(Continued)

Table 5.4 *(continued)*

Region/Country[2]	**2008**	**2015**	**2020**	**2025**	**2030**	**2035**	**% Change**
Mexico/Chile	1	1	1	1	2	2	2.4
OECD Europe	1	2	2	2	2	2	1.6
Japan	1	1	1	2	2	2	4.0
South Korea	0	0	0	0	0	0	0.0
Australia/New Zealand	1	1	2	2	2	2	4.7
Russia	0	0	0	0	0	0	4.3
China	0	0	0	0	0	0	—
India	0	0	0	0	0	0	—
Brazil	0	0	0	0	0	0	---
Other Eurasia	0	0	0	0	1	1	7.3
Other Asia	3	7	7	7	8	9	4.4
Other Central/South America	0	1	1	1	1	1	2.0
Middle East	0	0	0	0	0	0	—
Africa	0	0	0	0	0	0	4.6
WORLD	9	16	17	19	22	25	3.7
Solar Power							
Region/Country[2]	**2008**	**2015**	**2020**	**2025**	**2030**	**2035**	**% Change**
United States	1	9	11	11	12	13	8.8
Canada	0	0	0	0	0	0	7.7

Mexico/Chile	0	0	0	0	0	0	8.0
OECD Europe	10	32	35	37	38	40	5.4
Japan	2	8	12	16	20	27	9.8
South Korea	0	1	1	1	1	1	4.7
Australia/New Zealand	0	1	1	1	1	1	9.6
Russia	0	0	0	0	0	0	—
China	0	7	18	19	20	21	20.22
India	0	1	3	6	7	8	35.63
Brazil	0	0	0	0	0	0	—
Other Eurasia	0	0	0	0	0	0	22.3
Other Asia	0	0	0	0	0	0	13.7
Other Central/South America	0	0	0	0	0	0	12.3
Middle East	0	1	2	2	2	2	—
Africa	0	1	2	3	4	4	25.42
WORLD	14	62	86	97	106	119	8.3

[1] In gigawatts.
[2] Countries listed in bold are members of the Organisation for Economic Co-operation and Development (OECD). For more information about the OECD, see http://www.oecd.org/.

Source: International Energy Outlook 2011. Washington, DC: U.S. Energy Information Administration, Department of Energy. [September 19, 2011], Tables F7—F10. Available online at http://38.96.246.204/forecasts/ieo/pdf/0484(2011).pdf. Accessed March 10, 2012.

One of the environmental issues associated with the use of fossil fuels is the release of greenhouse gases to the atmosphere. The table below shows the trends in the role of fossil fuel combustion from various sources in the release of greenhouse gases (GHG) in the 27 member states of the European Union (EU 27) and the United States in selected years from 1990 to 2008, two regions of the world for which these data are reliably available.

Table 5.5 Sources of Greenhouse Gases, 1990–2008[1]

Source	1990	1995	2000	2005	2006	2007	2008
Energy							
EU (27)	3,489	3,188	3,047	3,095	3,076	2,999	2,945
US	5,288	6,168	6,283	6,210	6,291	6,117	5,751
Transportation							
EU (27)	778	836	915	968	974	979	962
US	1,545.2	—	1,932.3	2,017.4	1,994.4	2,003.8	1,890.7
Agriculture							
EU (27)	592	513	501	475	472	472	472
US	384	—	411	419	419	426	426
Industry							
EU (27)	484	463	413	420	421	434	410
US	316	—	349	334	339	351	332
Waste							
EU (27)	207	201	173	146	145	142	139
US	175	—	144	145	144	144	149
Solvent and other product use							
EU (27)	17	14	14	13	13	13	12
US	4.4	—	4.9	4.4	4.4	4.4	4.4

Sources: Eurostat. *Energy, Transport, and Environment Indicators.* Luxembourg: Publications Office of the European Union, 2011, Table 3.1.2, p. 140. Available online at http://epp.eurostat.ec.europa.eu/cache/ITY_OFFPUB/KS DK 10 001/EN/KS DK 10 001 EN.PDF; Accessed March 10, 2012. U.S. Environmental Protection Agency. *Inventory of U.S. Greenhouse Gases Emissions and Sinks: 1990–2009,* Table ES-4, p. ES-11, EPA 430-R-11-005. http://www.epa.gov/climatechange/emissions/downloads11/US GHG-Inventory-2011-Complete_Report.pdf. Accessed March 10, 2012.

A number of analysts have attempted to estimate the date on which a country will reach peak oil or, in some cases, has already reached peak oil. The following table shows the estimate by one group of researchers.

Table 5.6 Estimated Year in Which Peak Oil Will Occur or Has Occurred

Country	Estimated Peak Oil Year
Algeria	2012
Angola	2010
Indonesia	1977
Iran	1974
Iraq	2036
Kuwait	2033
Libya	2023
Nigeria	2017
Qatar	2019
Saudi Arabia	2027
United Arab Emirates	2030
Venezuela	2028
Argentina	1998
Australia	2000
Azerbaijan	2013
Brazil	2010
Brunei	1979
Canada	1973
China	2009
Columbia	1999
Congo	2000
Denmark	2004
Ecuador	2007
Egypt	1993
Equatorial Guinea	2011
Gabon	1996
India	2015
Italy	2006
Kazakhstan	2020
Malaysia	2004
Mexico	2004
Norway	2001
Oman	1998

(continued)

Table 5.6 (*continued*)

Country	Estimated Peak Oil Year
Peru	1985
Romania	1976
Russia	2009
Sudan	2015
Syria	1995
Thailand	2007
Trinidad and Tobago	1978
Tunisia	1984
Turkmenistan	2004
United Kingdom	1999
United States	1970
Uzbekistan	1995
Vietnam	2004
Yemen	2001
OPEC	2026
non-OPEC	2006
Total world	2014

Source: Sami Nashawi, Ibrahim, Adel Malallah, and Mohammed Al-Bisharah. "Forecasting World Crude Oil Production Using Multicyclic Hubbert Model." *Energy Fuels* 24, no. 3 (2010): 1796. Used by kind permission of the authors.

Documents

Political Consequences of the Status of World Energy Supplies (2005)

In late 2005, the Committee on Foreign Relations of the U.S. Senate invited former Secretary of Defense James R. Schlesinger to testify before its panel on "the complexity of U.S. reliance on imported energy sources, particularly oil, and the difficulties the U.S. faces in mediating detrimental effects of this dependency." The following is the gist of Schlesinger's remarks before the committee.

[Schlesinger begins by providing a brief historical review of the problems of energy security in Germany, France, Great Britain, and the United States. He then goes on to say that:]

Since [the 1960s], we have regularly talked about—and sought by various measures—to achieve greater energy security.

Such measures, limited as they were, have generally proved unsatisfactory. The nation's dependence on imported hydrocarbons has continued to surge.

Mr. Chairman, until such time as new technologies, barely on the horizon, can wean us from our dependence on oil and gas, we shall continue to be plagued by energy insecurity. We shall not end dependence on imported oil nor, what is the hope of some, end dependence on the volatile Middle East—with all the political and economic consequences that flow from that reality. That is not to say that various measures and inventions will not, from time to time, shave our growing dependence, but we will not end it. Instead of energy security, we shall have to acknowledge and to live with various degrees of insecurity.

To be sure, we have certain short-term problems to which I shall presently turn. More importantly, we face a fundamental, longer-term problem. In the decades ahead, we do not know precisely when, we shall reach a point, a plateau or peak, beyond which we shall be unable further to increase production of conventional oil worldwide. We need to understand that problem now and to begin to prepare for that transition.

[Schlesinger next discusses problems of supply and demand for petroleum products both in the United States and the rest of the world before turning to some social and political aspects of the world's energy crisis:]

Let me turn now to the political and economic ramifications. Again, let me underscore that energy actions tend to be a two-edged sword. To some extent, the recent higher prices for oil reflect some of our own prior policies and actions. For example, the sanctions imposed upon various rogue nations, by reducing world supply, have resulted in higher prices. Operation Iraqi Freedom, followed by the insurgency, has caused unrest in the Middle East. The consequence has been somewhat lower production and a significant risk premium that, again, has raised the price of oil.

The effect of higher oil prices has been significantly higher incomes for producers. A much higher level of income has meant that a range of nations, including Russia, Iran, Venezuela, as well as Gulf Arab nations have had their economic problems substantially eased. As a result, they have become less amenable to American policy initiatives. Perhaps more importantly, the flow of funds into the Middle East inevitably has added to the monies that can be transferred to terrorists. As long as the motivation is there and controls remain inadequate, that means that the terrorists will continue to be adequately or amply funded. To the extent that we begin to run into supply limitations and to the extent that we all grow more dependent on the Middle East, this problem of spillover funding benefits for terrorists is not going to go away.

There are, of course, additional problems of an economic nature. We all understand that higher oil prices can depress spending on other goods and services—and thereby cause slower growth rates and possibly a worldwide recession. The reverse side of rising receipts for producers is, of course, rising out-payments by consumer nations. This can readily augment structural imbalances. This year, the American balance-of-payments deficit looks to be almost three-quarters of a trillion dollars. That is not small change. Of the well over $700 billion of that deficit, some $300 billion comes from oil and gas. It is recognized that the U.S. balance-of-payments deficit represents the locomotive that drives much of the world's economies. In performing this service—for which we get little thanks—the United States is steadily adding to its financial obligations to others. How long this process can continue is uncertain, but high oil prices add to the dilemma.

Finally, Mr. Chairman, I must point to another problem. The United States is today the preponderant military power in the world. Still, our military establishment is heavily dependent upon oil. At a minimum, the rising oil price poses a budgetary problem for the Department of Defense at a time that our national budget is increasingly strained. Moreover, in the longer run, as we face the prospect of a plateau in which we are no longer able worldwide to increase the production of oil against presumably still rising demand, the question is whether the Department of Defense will still be able to obtain the supply of oil products necessary for maintaining our military preponderance.

In that prospective world, the Department of Defense will face all sorts of pressures at home and abroad to curtail its use of petroleum products, thereby endangering its overall military effectiveness.

Source: Statement of James Schlesinger before the Committee on Foreign Relations, United States Senate, November 16, 2005. Available online. http://lugar.senate.gov/energy/hearings/pdf/051116/Schlesinger_Testimony.pdf. Accessed March 5, 2012.

Peaking of World Oil Production: Impacts, Mitigation, and Risk Management (2005)

This report was conducted at the request of the U.S. Department of Energy, where lead researcher Robert L. Hirsch was project leader. The study was designed to determine the validity of the concept of oil peaking and to suggest some social, political, economic, and other consequences of such an event. Authors of the report drew the following major conclusions about oil peaking.

1. When world oil peaking will occur is not known with certainty. A fundamental problem in predicting oil peaking is the poor quality of and possible political biases in world oil reserves data. Some experts believe peaking may occur soon. This study indicates that "soon" is within 20 years.
2. The problems associated with world oil production peaking will not be temporary, and past "energy crisis" experience will provide relatively little guidance. The challenge of oil peaking deserves immediate, serious attention, if risks are to be fully understood and mitigation begun on a timely basis.

3. Oil peaking will create a severe liquid fuels problem for the transportation sector, not an "energy crisis" in the usual sense that term has been used.
4. Peaking will result in dramatically higher oil prices, which will cause protracted economic hardship in the United States and the world. However, the problems are not insoluble. Timely, aggressive mitigation initiatives addressing both the supply and the demand sides of the issue will be required.
5. In the developed nations, the problems will be especially serious. In the developing nations peaking problems have the potential to be much worse.
6. Mitigation will require a minimum of a decade of intense, expensive effort, because the scale of liquid fuels mitigation is inherently extremely large.
7. While greater end-use efficiency is essential, increased efficiency alone will be neither sufficient nor timely enough to solve the problem. Production of large amounts of substitute liquid fuels will be required. A number of commercial or near-commercial substitute fuel production technologies are currently available for deployment, so the production of vast amounts of substitute liquid fuels is feasible with existing technology.
8. Intervention by governments will be required, because the economic and social implications of oil peaking would otherwise be chaotic. The experiences of the 1970s and 1980s offer important guides as to government actions that are desirable and those that are undesirable, but the process will not be easy.

Source: Robert L. Hirsch, Roger Bezdek, and Robert Wendling. *Peaking of World Oil Production: Impacts, Mitigation, & Risk Management.* [n.p.], February 2005. Available online at http://www.netl.doe.gov/publications/others/pdf/oil_peaking_netl.pdf. Accessed March 5, 2012.

Crude Oil: Uncertainty about Future Oil Supply Makes It Important to Develop a Strategy for Addressing a Peak and Decline in Oil Production (2007)

In 2007 the U.S. Government Accountability Office conducted a study on peak oil in the United States and the world. It attempted to assess the timing of peak oil, the effects that event would have on the U.S. economy and the actions the U.S. government should take to deal with these effects. Following are the study's general results.

Results in Brief

Most studies estimate that oil production will peak sometime between now and 2040, although many of these projections cover a wide range of time, including two studies for which the range extends into the next century. The timing of the peak depends on multiple, uncertain factors that will influence how quickly the remaining oil is used, including the amount of oil still in the ground, how much of the remaining oil can be ultimately produced, and future oil demand. The amount of oil remaining in the ground is highly uncertain, in part because the Organization of Petroleum Exporting Countries (OPEC) controls most of the estimated world oil reserves, but its estimates of reserves are not verified by independent auditors. In addition, many parts of the world have not yet been fully explored for oil. There is also great uncertainty about the amount of oil that will ultimately be produced, given the technological, cost, and environmental challenges. For example, some of the oil remaining in the ground can be accessed only by using complex and costly technologies that present greater environmental challenges than the technologies used for most of the oil produced to date. Other important sources of uncertainty about future oil production are potentially unfavorable political and investment conditions in countries where oil is located. For example, more than 60 percent of world oil reserves, on the basis of Oil and Gas Journal estimates, are in countries where relatively unstable political conditions could constrain oil exploration and production. Finally, future world demand for oil also is uncertain because it depends on economic growth and government policies throughout the world. For example, continued rapid economic growth in China and India could significantly increase world demand for oil, while environmental concerns, including oil's contribution to global warming, may spur conservation or adoption of alternative fuels that would reduce future demand for oil.

In the United States, alternative transportation technologies face challenges that could impede their ability to mitigate the consequences of a peak and decline in oil production, unless sufficient time and effort are brought to bear. For example:

- Ethanol from corn is more costly to produce than gasoline, in part because of the high cost of the corn feedstock. Even if ethanol were to become more cost-competitive with gasoline, it could not become widely available without costly investments in infrastructure, including pipelines, storage tanks, and filling stations.
- Advanced vehicle technologies that could increase mileage or use different fuels are generally more costly than conventional technologies and have not been widely adopted. For example, hybrid electric vehicles can cost from $2,000 to $3,500 more to purchase than comparable conventional vehicles and currently constitute about 1 percent of new vehicle registrations in the United States.
- Hydrogen fuel cell vehicles are significantly more costly than conventional vehicles to produce. Specifically, the hydrogen fuel cell stack needed to power a vehicle currently costs about $35,000 to produce, in comparison with a conventional gas engine, which costs $2,000 to $3,000.

Given these challenges, development and widespread adoption of alternative transportation technologies will take time and effort. Key alternative technologies currently supply the equivalent of only about 1 percent of U.S. consumption of petroleum products, and DOE projects that even under optimistic scenarios, by 2015 these technologies could displace only the equivalent of 4 percent of projected U.S. annual consumption. Under these circumstances, an imminent peak and sharp decline in oil production could have severe consequences, including a worldwide recession. If the peak comes later, however, these technologies have a greater potential to mitigate the consequences. DOE projects that these technologies could displace up to the equivalent of 34 percent of projected U.S. annual consumption of petroleum products in the 2025 through 2030 time frame, assuming the challenges the technologies face are overcome. The level of effort dedicated to overcoming challenges to alternative technologies will depend in part on the price of oil; without sustained high oil prices, efforts to develop and adopt alternatives may fall by the wayside.

Federal agency efforts that could reduce uncertainty about the timing of peak oil production or mitigate its consequences are spread across multiple agencies and generally are not focused explicitly on peak oil. For example, efforts that could be used to reduce uncertainty about the timing of a peak include USGS activities to estimate oil resources and DOE efforts to monitor current supply and demand conditions in global oil markets and to make future projections. Similarly, DOE, the Department of Transportation (DOT), and the U.S. Department of Agriculture (USDA) all have programs and activities that oversee or promote alternative transportation technologies that could mitigate the consequences of a peak. However, officials of key agencies we spoke with acknowledge that their efforts—with the exception of some studies—are not specifically designed to address peak oil. Federally sponsored studies we reviewed have expressed a growing concern over the potential for a peak and officials from key agencies have identified some options for addressing this issue. For example, DOE and USGS officials told us that developing better information about worldwide demand and supply and improving global estimates for nonconventional oil resources and oil in "frontier" regions that have yet to be fully explored could help prepare for a peak in oil production by reducing uncertainty about its timing. Agency officials also said that, in the event of an imminent peak, they could step up efforts to mitigate the consequences by, for example, further encouraging development and adoption of alternative fuels and advanced vehicle technologies. However, according to DOE, there is no formal strategy for coordinating and prioritizing federal efforts dealing with peak oil issues, either within DOE or between DOE and other key agencies.

While the consequences of a peak would be felt globally, the United States, as the largest consumer of oil and one of the nations most heavily dependent on oil for transportation, may be particularly vulnerable. Therefore, to better prepare the United States for a peak and decline in oil production, we are recommending that the Secretary of Energy take the lead, in coordination with other relevant federal agencies, to establish a peak oil strategy. Such a strategy should include efforts to reduce uncertainty about the timing of a peak in oil production and provide timely advice to Congress about cost-effective measures to mitigate the potential consequences of a peak. In commenting on a draft of the report, the Departments of Energy and the Interior generally agreed with the report and recommendations.

Source: U.S. Government Accountability Office. *Crude Oil: Uncertainty about Future Oil Supply Makes It Important to Develop a Strategy for Addressing a Peak and Decline in Oil Production.* Washington, D.C.: Government Accountability Office, February 2007, 4–6.

2050: The Future Begins Today—Recommendations for the EU's Future Integrated Policy on Climate Change (2009)

Various nations have taken different approaches to the problem of climate change with varying degrees of enthusiasm. Almost since the moment of signing of the Kyoto Protocol, the European Union has led the world in thinking about ways in which the union could meet the goals laid out in Kyoto. After extended discussions and research, the European Parliament voted in February 2009 on an overarching policy on climate change for the EU. The final document was very long, and the section excerpted here outlines only some of the most important guiding principles behind the final document. The document has been edited to include only the core ideas of each section included here. Interested readers should examine the complete document to best understand the many detailed plans that have been proposed based on the general philosophy elucidated here.

Guiding Political Ideas

1. . . . all efforts to curb emissions should aim at staying well below the objective of limiting global temperature increases to below 2°C, inasmuch as a level of warming of that magnitude would already impact heavily on our society and individual lifestyles and would also entail significant changes in ecosystems and water resources; . . . climate change is both more rapid and more serious in terms of its adverse effects than was previously thought; consequently, [the Parliament] calls on the Commission to closely monitor and analyse the latest scientific findings with a view to assessing, in particular, whether the EU 2°C target would still achieve the aim of avoiding dangerous climate change;
2. . . . there is an urgent need . . . to incorporate global warming and climate change as new parameters into all spheres and policies, and to take the causes and consequences of global warming and climate change into account in every relevant area of EU legislation;
3. . . . a medium-term target of a 25%–40% reduction in greenhouse gas emissions by 2020, as well as a long-term reduction target of at least 80% by 2050, compared to 1990 . . . ;

4. . . . urges the EU to take urgent steps at home and in the context of international negotiations to develop accounting principles that also include the full effects of consumption, including the effects of international aviation;
5. Calls on the Commission to consider the carbon footprints of future European policy initiatives so as to ensure that climate change targets set at European level are met, whilst still ensuring a high level of protection for the environment and public health;
6. Stresses the political measures, and cooperation at international level (including regional multilateral agreements) and at EU and Member State level, repeatedly proposed by Parliament with a view to combating climate change;
7. Welcomes the adoption of the EU package of legislative measures (the so-called "climate and energy package") requiring the unilateral reduction by 20% of EU greenhouse gas emissions, setting up a procedure to step up the effort to achieve a 30% reduction in accordance with commitments under the future international agreement and increasing to 20% the share of renewable energy in the EU energy mix by 2020, and calls on the EU Member States to implement those legislative measures smoothly and rapidly; calls on the Commission to monitor the implementation of the "climate and energy package" closely; . . .
11. Urges the Commission and the Member States to support the UN's call for a "Green New Deal"; in the light of the financial crisis, calls for the investments aimed at boosting economic growth to do so in a sustainable way, in particular by promoting green technologies which will at the same time advance Europe's future competitiveness and secure jobs;
12. Emphasises *[sic]*, in this context, that tackling climate change will lead to societal changes that will help to create new jobs and industries, combat energy poverty and

dependency on imports of fossil fuels and provide social benefits for citizens; stresses that cooperation at international, regional and local level will be critical if we are to be successful in achieving this goal;

13. Is convinced, moreover, that climate change can only be successfully combated if citizens are fully engaged in the process and are protected during the period of transition to a carbon-neutral economy; highlights, therefore, the fact that mitigation and adaptation policies will push the European Union towards a new model of sustainable development which should promote its social character in order to secure the social consensus;
14. Stresses the need, first of all, to achieve dramatic improvements in efficiency in all areas of everyday life and, in parallel, to launch a sustainable production and consumption model with a conscious saving of resources on the basis of renewable energy;
15. Emphasises in this context the need to examine the EU's budget, and existing and future financing instruments, as to their compatibility with European climate policy, and where necessary to adapt them;
16. Stresses that a successful R&D policy will only be made possible by the practical application of new technologies via secured market access points;
17. Calls for research to be carried out into potential trends of climate-induced migration and the ensuing pressures on local services, in order to inform long-term planning and risk-management processes; . . .

[The resolution continues with a total of more than 200 specific recommendations in areas such as Energy, Biofuels, Energy Efficiency, Mobility and Logistics, Industrial Emissions, Agriculture and Livestock Breeding, Forests, Soil Protection, Water Management, Fisheries, Waste Treatment and Resource Management, and Adaptation Measures. The resolution concludes with the following sections:]

202. Calls for an agenda for action to combat climate change for the period 2009–2014, to be implemented as follows:

a) at EU level, the Commission and the Member States should:

   – lead discussions at a local and global level on actions to be taken to combat climate change,
   – develop, fund and introduce an EU-wide supergrid accessible to all forms of electricity providers,
   – promote and fund efficient, sustainable transport infrastructure to reduce carbon emissions, including hydrogen technology and high-speed railways,
   – develop new communication strategies to educate citizens and provide them with incentives to reduce emissions in an affordable way, e.g. by developing information on the carbon content of products and services,
   – develop appropriate legislative instruments to encourage all industrial sectors to become leaders in the fight against climate change, starting with a demand for transparency on carbon emissions,
   – establish stronger links between the Lisbon policy agenda, the social agenda and climate change policies;

b) at local and regional level, best practices should be promoted and exchanged, in particular concerning:

   – energy efficiency and other measures to combat energy poverty, with the objective of net-zero-energy performance targets in private, commercial and public buildings,
   – the recycling and re-utilisation of waste, for instance by developing infrastructures for collection points,
   – the development of infrastructures for low-emission passenger cars using renewable energies, as well as the introduction of incentives for the development of zero-emission vehicles for public transport,

- the promotion of more sustainable mobility in cities and in rural areas,
- the adoption and implementation of measures for adaptation to climate change,
- the promotion of local and regional food production and consumption[.]

Source: *2050: the Future Begins Today—Recommendations for the EU's Future Integrated Policy on Climate Change.* Resolution adopted by the European Parliament, 4 February 2009. Available online. http://www.europarl.europa.eu/sides/getDoc.do?type = TA&reference = P6-TA-2009-0042&language = EN#BKMD-1. Accessed March 7, 2012.

International Energy Outlook (2011)

One of the provisions of the Department of Energy Organization Act of 1977 was that the Department's Energy Information Administration (EIA) annually prepare a report projecting energy production and consumption over an extended period into the future. In September 2011, the EIA released its report outlining its best projections for world energy production and use into the year 2035. Some of the major findings in that report are provided here. The abbreviation IEO2011 stands for International Energy Outlook 2011, and the term "reference case" refers to a method by which data from a variety of sources is compared in order to make some useful generalization or projection about the future. Figure references have been omitted from the selection.

In the long term, the IEO2011 Reference case projects increased world consumption of marketed energy from all fuel sources through 2035. Fossil fuels are expected to continue supplying much of the energy used worldwide. Although liquid fuels—mostly petroleum based—remain the largest source of energy, the liquids share of world marketed energy consumption falls from 34 percent in 2008 to 29 percent in 2035, as projected high world oil prices lead many energy users to switch away from liquid fuels when feasible. Renewable energy is the world's fastest growing form of energy, and the renewable share of total energy use increases from 10 percent in 2008 to 14 percent in 2035 in the Reference case.

Liquid Fuels

World use of petroleum and other liquids grows from 85.7 million barrels per day in 2008 to 97.6 million barrels per day in 2020 and 112.2 million barrels per day in 2035. In the Reference case, most of the growth in liquids use is in the transportation sector, where, in the absence of significant technological advances, liquids continue to provide much of the energy consumed. Liquid fuels remain an important energy source for transportation and industrial sector processes. Despite rising fuel prices, use of liquids for transportation increases by an average of 1.4 percent per year, or 46 percent overall from 2008 to 2035. The transportation sector accounts for 82 percent of the total increase in liquid fuel use from 2008 to 2035, with the remaining portion of the growth attributable to the industrial sector. The use of liquids declines in the other end-use sectors and for electric power generation.

To meet the increase in world demand in the Reference case, liquids production (including both conventional and unconventional liquids supplies) increases by a total of 26.6 million barrels per day from 2008 to 2035. The Reference case assumes that OPEC countries will invest in incremental production capacity in order to maintain a share of approximately 40 percent of total world liquids production through 2035, consistent with their share over the past 15 years. Increasing volumes of conventional liquids (crude oil and lease condensate, natural gas plant liquids, and refinery gain) from OPEC producers contribute 10.3 million barrels per day to the total increase in world liquids production, and conventional supplies from non-OPEC countries add another 7.1 million barrels per day.

Unconventional resources (including oil sands, extra-heavy oil, biofuels, coal-to-liquids, gas-to-liquids, and shale oil) from both OPEC and non-OPEC sources grow on average by 4.6 percent per year over the projection period. Sustained high oil prices allow unconventional resources to become economically competitive, particularly when geopolitical or other "above ground" constraints limit access to prospective conventional resources. World production of unconventional liquid fuels, which totaled only 3.9 million barrels per day in 2008, increases to 13.1 million barrels per day and accounts for 12 percent of total world liquids supply in 2035. The largest components of future unconventional production are 4.8 million barrels per day of Canadian oil sands, 2.2 and 1.7 million barrels per day of U.S. and Brazilian biofuels, respectively, and 1.4 million barrels per day of Venezuelan extra-heavy oil. Those four contributors to unconventional liquids supply account for almost three-quarters of the increase over the projection period.

Natural Gas

World natural gas consumption increases by 52 percent in the Reference case, from 111 trillion cubic feet in 2008 to 169 trillion cubic feet in 2035. Although the global recession resulted in an estimated decline of 2.0 trillion cubic feet in natural gas use in 2009, robust demand returned in 2010, and consumption exceeded the level recorded before the downturn. Natural gas continues to be the fuel of choice for many regions of the world in the electric power and industrial sectors, in part because its relatively low carbon intensity compared with oil and coal makes it an attractive option for nations interested in reducing greenhouse gas emissions. In the power sector, low capital costs and fuel efficiency also favor natural gas.

In the IEO2011 Reference case, the major projected increase in natural gas production occurs in non-OECD regions, with the largest increments coming from the Middle East (an increase of 15 trillion cubic feet between 2008 and 2035), Africa (7 trillion cubic feet), and non-OECD Europe and Eurasia, including Russia and the other former Soviet Republics (9 trillion cubic feet). Over the projection period, Iran and Qatar alone increase their natural gas production by a combined 11 trillion cubic feet, nearly 20 percent of the total increment in world gas production. A significant share of the increase is expected to come from a single offshore field, which is called North Field on the Qatari side and South Pars on the Iranian side.

Contributing to the strong competitive position of natural gas among other energy sources is a strong growth outlook for reserves and supplies. Significant changes in natural gas supplies and global markets occur with the expansion of liquefied natural gas (LNG) production capacity and as new drilling techniques and other efficiencies make production from many shale basins economical worldwide. The net impact is a significant increase in resource availability, which contributes to lower prices and higher demand for natural gas in the projection.

Although the extent of the world's unconventional natural gas resources—tight gas, shale gas, and coalbed methane—have not yet been assessed fully, the IEO2011 Reference case projects a substantial increase in those supplies, especially from the United States but also from Canada and China. An initial assessment of shale gas resources in 32 countries was released by EIA in April 2011. The report found that technically recoverable shale gas resources in the assessed shale gas basins and the United States amount to 6,622 trillion cubic feet. To put the shale gas resource estimate in perspective, according to the Oil & Gas Journal world proven reserves of natural gas as of January 1, 2011, are about 6,675 trillion cubic feet, and world technically recoverable gas resources—largely excluding shale gas—are roughly 16,000 trillion cubic feet.

Rising estimates of shale gas resources have helped to increase total U.S. natural gas reserves by almost 50 percent over the past decade, and shale gas rises to 47 percent of U.S. natural gas production in 2035 in the IEO2011 Reference case. Adding production of tight gas and coalbed methane, U.S. unconventional natural gas production rises from 10.9 trillion cubic feet in 2008 to 19.8 trillion cubic feet in 2035. Unconventional natural gas resources are even more important for the future of domestic gas supplies in Canada and China, where they account for 50 percent and 72 percent of total domestic production, respectively, in 2035 in the Reference case.

World natural gas trade, both by pipeline and by shipment in the form of LNG, is poised to increase in the future. Most of the projected increase in LNG supply comes from the Middle East and Australia, where a number of new liquefaction projects are expected to become operational within the next decade. Additionally, several LNG export projects have been proposed for western Canada, and there are also proposals to convert underutilized LNG import facilities in the United States to liquefaction and export facilities for domestically sourced natural gas. In the IEO2011 Reference case, world liquefaction capacity more than doubles, from about 8 trillion cubic feet in 2008 to 19 trillion cubic feet in 2035. In addition, new pipelines currently under construction or planned will increase natural gas exports from Africa to European markets and from Eurasia to China.

Coal

In the absence of national policies and/or binding international agreements that would limit or reduce greenhouse gas emissions, world coal consumption is projected to increase from 139 quadrillion Btu in 2008 to 209 quadrillion Btu in 2035, at an average annual rate of 1.5 percent. Regional growth rates are uneven, with little growth in coal consumption in OECD nations but robust growth in non-OECD nations, particularly among the Asian economies.

Strong economic growth and large domestic coal reserves in China and India lead to a substantial increase in their coal use for electric power and industrial processes. Installed coal-fired generating capacity in China nearly doubles in the Reference case from 2008 to 2035, and coal use in China's industrial sector grows by 67 percent. The development of China's electric power and industrial sectors will require not only large-scale infrastructure investments but also substantial investment in both coal mining and coal transportation infrastructure. In India, coal-fired generating capacity rises from 99 gigawatts in 2008 to 172 gigawatts in 2035, a 72-percent increase, while industrial sector coal use grows by 94 percent.

Source: *International Energy Outlook 2011.* Washington, DC: U.S. Energy Information Administration, Department of Energy. [September 19, 2011], pp. 1–4. Available online. http://38.96.246.204/forecasts/ieo/pdf/0484(2011).pdf. Accessed March 9, 2012.

H.R. 1868. Clean Coal-Derived Fuels for Energy Security Act of 2011 (2011)

Legislative bodies quite naturally attempt to influence the direction of energy development in their nations through the adoption of laws and regulations that provide support for, favor, or discourage one type of energy over others. The bill summarized here is only one of many such legislative actions that appear in legislatures all over the world every year. At the time this book has gone to press, no action has been taken on this bill.
Section I of the bill provides its short title, as given above.
Section II provides definitions of terms used in the bill.

SEC. 3. CLEAN COAL-DERIVED FUEL PROGRAM.

a) PROGRAM.—

(1) IN GENERAL.—Not later than 1 year after the date of enactment of this Act, the President shall promulgate regulations to ensure that covered fuel sold or introduced into commerce in the United States (except in noncontiguous States or territories), on an annual average basis, contains the applicable volume of clean coal-derived fuel determined in accordance with paragraph (4).

(2) PROVISIONS OF REGULATIONS.—The regulations promulgated under paragraph (1)—

(A) shall contain compliance provisions applicable to refineries, blenders, distributors, and importers, as appropriate, to ensure that—

(i) the requirements of this subsection are met; and

(ii) clean coal-derived fuels produced from facilities for the purpose of compliance with this Act result in lifecycle green house gas emissions that are not greater than gasoline; and

(B) shall not—

(i) restrict geographic areas in the contiguous United States in which clean coal-derived fuel may be used; or

(ii) impose any per-gallon obligation for the use of clean coal-derived fuel.

(3) RELATIONSHIP TO OTHER REGULATIONS.—Regulations promulgated under this subsection shall, to the extent practicable, incorporate the program structure and compliance and reporting requirements established under the final regulations promulgated to implement the renewable fuel program under section 211(o) of the Clean Air Act (42 U.S.C. 7545(o)).

(4) APPLICABLE VOLUME.—

(A) CALENDAR YEARS 2017 THROUGH 2024.—For the purpose of this subsection, the applicable volume for any of calendar years 2017 through 2024 shall be determined in accordance with the following table:

Calendar year	Applicable volume of clean coal-derived fuel: (in billions of gallons)
2017	0.75
2018	1.5
2019	2.25
2020	3.00
2021	3.75
2022	4.5
2023	5.25
2024	6.0.

(B) CALENDAR YEAR 2025 AND THEREAFTER.—Subject to subparagraph (C), for the purposes of this subsection, the applicable volume for calendar year 2025 and each calendar year thereafter shall be determined by the President, in coordination with the Secretary and the Administrator of the Environmental Protection Agency, based on a review of the implementation of the program during calendar years 2017 through 2024, including a review of—

(i) the impact of clean coal-derived fuels on the energy security of the United States;

(ii) the expected annual rate of future production of clean coal-derived fuels; and

(iii) the impact of the use of clean coal-derived fuels on other factors, including job creation, rural economic development, and the environment.

(C) MINIMUM APPLICABLE VOLUME.—For the purpose of this subsection, the minimum applicable volume for calendar year 2025 and each calendar year thereafter shall be equal to the product obtained by multiplying—

(i) the number of gallons of covered fuel that the President estimates will be sold or introduced into commerce in the calendar year; and

(ii) the ratio that—

(I) 6,000,000,000 gallons of clean coal-derived fuel; bears to

(II) the number of gallons of covered fuel sold or introduced into commerce in calendar year 2024.

[The bill then continues with more details as to how the amount of liquid fuels produced is to be calculated, credit programs associated with the programs, special waivers, and other details about the execution of the final act.]

Source: H.R. 1868. http://frwebgate.access.gpo.gov/cgi-bin/getdoc.cgi?dbname = 112_cong_bills&docid = f:h1868ih.txt.pdf. Accessed November 20, 2011.

CAT
D8T

Issues related to the production and consumption of energy resources have been of worldwide concern for at least half a century. During that time, countless numbers of books, articles, reports, Internet postings, and other print and nonprint materials have been published on the world's energy crisis. Few bibliographies can begin to review the many resources that are currently available on this topic. This chapter attempts to list a sampling of important print and nonprint resources dealing with worldwide energy issues. The chapter is divided into two sections, the first listing print resources, and the second, nonprint resources. The former is divided into three subsections: books, periodicals, and reports. The latter consists almost entirely of Internet postings. Of course, a large number of magazine and journal articles have also been written on the subject of global energy issues. In many cases, these articles are also available on the Internet. Such articles are, therefore, included in the nonprint section for Internet listings.

Print Resources

Books

Andrews-Speed, Philip, and Roland Dannreuther. *China, Oil and Global Politics.* London: Routledge, 2011.

A bulldozer pushes clean coal into piles at the Mettiki Maryland Coal Mine. (AP Photo/Chris Gardner, File)

In the first half of this book, the authors explore the development of energy production and consumption, along with related energy policy in China over the past few decades. They then review some of the ways in which domestic trends and policies have had and are likely to continue to have far-reaching effects on the rest of the world.

Ayres, Robert U., and Edward H. Ayres. *Crossing the Energy Divide: Moving from Fossil Fuel Dependence to a Clean-energy Future.* Upper Saddle River, NJ: Wharton School Publishing, 2010.

The authors make two essential points in this book. First, energy plays a far more fundamental role in the operation of modern society than most economists acknowledge and second, the world's reliance on fossil fuels is so profound that it will take decades to convert from oil, gas, and coal to alternative forms of energy. They suggest some ways in which that transition can occur.

Banerjee, Sudeshna Ghosh, Avjeet Singh, and Hussain A. Samad. *Power and People: The Benefits of Renewable Energy in Nepal.* Washington, DC: World Bank Publications, 2011.

This World Bank report was issued to Nepal's Alternative Energy Promotion Center, outlining the possible role of alternative energy in the nation's future energy policy mix. It provides an interesting insight into the types of energy issues that nations of all sizes and economic power face in a world of decreasing fossil fuel supplies.

Brune, Michael. *Coming Clean: Breaking America's Addiction to Oil and Coal,* 2nd ed. Berkeley, CA: Counterpoint Press, 2010.

Brune is a political activist and currently (2011) director of the Rainforest Action Network. He argues that the majority of people are supportive of aggressive actions to reduce our dependence on fossil fuels, but that corporate and political interests are a powerful factor in preventing the needed transition to alternative forms of energy. Some of the topics covered in the book are "The Dirty Side of 'Clean Coal,'" "The True Cost of America's Oil Habit," "Moving America in a Post-Oil Economy," and "How Sun and Wind Can Generate Power and Prosperity."

Bryce, Robert. *Power Hungry: The Myths of "Green" Energy and the Real Fuels of the Future.* New York: PublicAffairs, 2010.

The author takes the somewhat contrarian view that the energy problem facing the world today is not that people are using too much fossil fuels, but too little. He argues that calls for increasing our use of alternative energy sources are misguided in that such resources will never be able to meet the demands of a modern society. He suggests that the best realistic option for meeting near-term energy demands is natural gas, which is relatively abundant, inexpensive, and environmentally harmless.

Carbonnier, Gilles, ed. *International Development Policy: Energy and Development.* London: Palgrave Macmillan, 2011.

This collection of essays deals with energy-related issues in a variety of countries and regions around the world, including sub-Saharan Africa, Eastern Europe, Central Asia, Ecuador, China, and India.

Carollo, Salvatore. *Understanding Oil Prices: A Guide to What Drives the Price of Oil in Today's Markets,* 2nd ed. New York: Wiley, 2012.

The author attempts to demystify the seemingly random fluctuations in oil prices over the past half century by explaining the factors that affect such changes. The book is an excellent general introduction to a difficult economic issue.

Chevalier, Jean-Marie, ed. *The New Energy Crisis: Climate, Economics and Geopolitics.* Houndmills, UK: Palgrave Macmillan, 2012.

This collection of articles reviews the world's current energy crisis from a number of perspectives, including the carbon-dependent Asian economies, the curse or blessing of oil production for Middle East and African states, energy poverty and economic development, and the energy crisis as a turning point for American society.

Crane, Hewitt D., Edwin Max Kinderman, and Ripudaman Malhotra. *A Cubic Mile of Oil: Realities and Options for Averting the Looming Global Energy Crisis.* Oxford: Oxford University Press, 2010.

The cubic mile referred to in the title is the approximate amount of oil used by the world annually. The authors review the process by which the world has begun to approach an international energy crisis and some of the ways that crisis can be resolved.

Darley, Julian. *High Noon for Natural Gas: The New Energy Crisis.* White River Junction, VT: Chelsea Green Publishing, 2004.

To many observers, increased use of natural gas is the best short-term solution to many world energy problems. It is thought to be abundant, relatively inexpensive, and environmentally more benign that coal or petroleum. In this book, however, the author argues that natural gas may not be the savior for world energy problems that many have hoped for it to be. He argues that our increasing dependence on natural gas may be the nation's (and the world's) next energy crisis.

Deffeyes, Kenneth S. *Beyond Oil: The View from Hubbert's Peak.* New York: Hill and Wang, 2006.

The major theme of this book is a consideration of the peak period for petroleum production, mathematically determined by M. King Hubbert in the last half of the 20th century. The author explains Hubbert's research, adds a discussion of his own extension of that research, and describes the effects of its conclusion on energy policy for the future. The last half of the book is devoted to a review of possible alternatives to petroleum as a world energy resource, including tar sands, oil shale, natural gas, uranium, and hydrogen.

Deutsch, John M. *The Crisis in Energy Policy.* Cambridge, MA: Harvard University Press, 2011.

This book is based on the author's Godkin Lectures on the Essentials of Free Government and the Duties of the Citizen, delivered at Harvard University in 2010. Deutsch argues that energy policy is based on three goals: learning how to deal with the problem of climate change; transitioning from fossil fuels to renewable energy technologies; and decreasing our dependency on imported oil by learning how to use existing energy sources more efficiently. He suggests how the United States has fallen short in working toward those goals, and how it can do better in the future.

Downey, Morgan. *Oil 101.* New York: Wooden Table Press, 2009.

Downey is a commodities trader said to be an expert in petroleum stocks. He has written a book that provides a superb introduction to the economics of petroleum production and consumption. Chapters deal with topics such as a brief history of oil, the physical and chemical composition of petroleum, petrochemicals, transportation of petroleum and petroleum products, environmental issues, pricing policies, and future prospects for the petroleum market.

Ebinger, Charles K. *Energy and Security in South Asia: Cooperation or Conflict?* Washington, DC: Brookings Institution Press, 2011.

The author considers the special energy issues faced by the nations of South Asia arising out of their rapid economic growth, the increasing costs of their dependence on nations of the Middle East for their supplies of petroleum, the expanded dependence on coal in some parts of the region, and the effects of political conflicts among nations in the region. He explores some of the ways in which South Asian nations can deal with the growing energy crisis, including the potential for greater use of alternative forms of energy.

Editors of *Scientific American* Magazine. *Oil and the Future of Energy.* Guilford, CT: The Lyons Press, 2007.

This collection of essays attempts to cover the field of energy issues facing the world today. They include an overview of the production and consumption of fossil fuels, the geopolitics related to the petroleum industry, potential environmental consequences of our dependence on fossil fuels (such as global climate change), options available for the use of fossil fuels, and strategies for the transition from a fossil-fuel-based to an alternative-fuel-based world economy.

Freese, Barbara. *Coal: A Human History.* New York: Penguin, 2004.

The author presents a comprehensive history of the use of coal as an energy resource, dating back to the early Chinese and Roman periods. She discusses current issues associated with its contemporary use as a major energy source for human activities, with special attention to the economic and political aspects of its use. She also reviews some of the harmful effects of coal-based economies on the men and women who work in the industry.

Geri, Laurance R., and David E. McNabb. *Energy Policy in the U.S.: Politics, Challenges, and Prospects for Change.* Boca Raton, FL: CRC Press, 2011.

The authors divide their book into two parts, the first of which discusses the history of energy policy in the United States and important individuals and organizations involved in that history. The second part of the book reviews specific kinds of policy decisions that affect energy production and consumption and ways in which they might be implemented to produce specific outcomes in the future.

Goodell, Jeff. *Big Coal: The Dirty Secret Behind America's Energy Future.* Boston: Mariner Books, 2007.

Goodell focuses on the role of coal in the American economy, pointing out that it is the "elephant in the room" with regard to the nation's (and the world's) energy future. As supplies of petroleum continue to run out, there is greater emphasis on the use of coal as a substitute for its sister fossil fuel. The author discusses in some detail the political and economic power of coal companies in the United States, and focuses on the deplorable working conditions that coal workers are forced to tolerant. He completes the book with a review of some ways in which the nation can avoid a future in which coal plays a dominant role in its energy policies.

Goodstein, David. *Out of Gas: The End of the Age of Oil.* New York: W. W. Norton, 2005.

A number of authors have written apocalyptic warnings about the coming end of the age of fossil fuels. This book is a good example of such works. Goodstein argues that the world may exhaust its supply of petroleum reserves by the year 2015, and is probably already past King Hubbert's so-called peak point, at which half of the world's petroleum supplies have already been exhausted. The author points to nuclear power as one of the few available options for dealing with world energy needs once fossil fuels have been depleted.

Gorelick, Steven M. *Oil Panic and the Global Crisis: Predictions and Myths.* Hoboken, NJ: Wiley-Blackwell, 2010.

Gorelick lays out the assumptions that lie at the current concerns about a worldwide energy crisis, then proceeds to explain why some of those assumptions are incorrect or inaccurate, and how this new view of the crisis should change people's view of the current world energy situation.

Graetz, Michael J. *The End of Energy: The Unmaking of America's Environment, Security, and Independence.* Cambridge, MA: MIT Press, 2011.

The author provides a thorough review of the status and potential impact of the energy crisis in the United States discussing the role of coal, oil, gas, and nuclear energy; the confidence crisis in the nation; the effects of climate change on the debate over energy use; and the role of government and economics on changes in patterns of energy use in the country.

Greer, John Michael. *The Long Descent: A User's Guide to the End of the Industrial Age.* Gabriola Island, BC: New Society Publishers, 2008.

The dual threat of global climate change and diminishing oil supplies has produced a number of apocalyptic books that warn of the massive collapse of human civilization. Greer's tome is one of these. He argues that the world has started down a long and arduous path that will result in the loss of many of the accomplishments made possible by the long use of fossil fuels. Since governments appear unable to find ways of dealing with this transformation by adopting new ways of thinking about and acting on these issues, the only hope for human survival appears to be the individuals take greater command of their own lives.

Hallowes, David. *Toxic Futures: South Africa in the Crises of Energy, Environment and Capital.* Scottsville, South Africa: University of KwaZulu-Natal Press, 2011.

The author reviews the current world energy crisis, including the growing threat of climate change, and concludes that the crisis has developed as a result of the "collusion between state and corporate power" that has resulted in a "war on the poor and on the environment." He presents a fascinating analysis of the world energy crisis that differs very much from what one reads in books written by authors from more developed nations.

Heinberg, Richard. *Blackout: Coal, Climate and the Last Energy Crisis.* Forest Row, East Sussex, UK: Clairview Books, 2009.

Coal is already an important factor in the world's energy equation, accounting for about half of all the electricity produced in the United States, and more than 70 percent of the energy used in China. Many observers have suggested that the inability of petroleum resources to keep up with growing worldwide demand means that coal will become an even more important energy source in the next century. However, coal has some serious drawbacks itself. It is, in the first place, the dirtiest of all fossil fuels, accounting for the release of very significant amounts of carbon dioxide to the Earth's atmosphere. Second, the cost of transporting may be increasing at a faster rate than its decreasing cost of production. Perhaps of greatest concern are recent studies that suggest that supplies of coal may not be as great as are sometimes estimated, and that the world may actually be facing a coal peak within a matter of decades.

Heinberg, Richard. *The Party's Over: Oil, War and the Fate of Industrial Societies,* 2nd ed. Gabriola Island, BC: New Society Publishers, 2005.

The author has written extensively on the coming social and economic crisis posed by reducing supplies of petroleum and other fossil fuels. In this book, Heinberg provides a general review of the world's energy crisis, beginning with an excellent review of the history of energy use in human civilization, the rise of the dominance of fossil fuels, the current status of energy resources, and adjustments that will be necessary once fossil fuels are no longer as abundant as they are today.

Holmgren, David. *Future Scenarios: How Communities Can Adapt to Peak Oil and Climate.* White River Junction, VT: Chelsea Green Publishing, 2009.

Holmgren explores four possible future scenarios, resulting from either mild or catastrophic climate change combined with either slow or rapid depletion of fossil fuel supplies. Each scenario explores possible social, political, agricultural, and technological responses to the challenges posed by each possible pair of climate and energy changes. The scenarios are fleshed out with reference to specific societies in existence today that have evolved to meet a specific set of demands and opportunities.

Hopkins, Rob. *The Transition Handbook: From Oil Dependency to Local Resilience.* White River Junction, VT: Chelsea Green Publishing, 2008.

The author, cofounder of the Transition Network, argues that the twin threat of global climate change and decreasing energy supplies inevitably means that dramatic changes will occur in the way people and communities live their lives. The key concept, he says, can only be that "small is beautiful." He reviews the development of this concept in a number of cities and towns in the Transition Towns movement in the United Kingdom, where people are finding new ways to solve everyday problems of transportation, food, building materials, and waste management.

Ibarrarán, Maria Eugenia, and Roy Boyd. *Hacia el Futuro: Energy, Economics and the Environment in 21st Century Mexico.* New York: Springer, 2011. Reprint of 2006 edition.

The authors discuss the potential effects of energy policy in Mexico on the economic growth, use of fossil fuel resources, and environmental effects over the next two decades. They make comparisons in these projections with some of Mexico's neighbors in Central and South America.

Klare, Michael T. *Rising Powers, Shrinking Planet: The New Geopolitics of Energy.* New York: Holt Paperbacks, 2009.

Klare claims that decreasing energy supplies will bring about a dramatic reformulation of geopolitical power throughout the world, with Russia being the short-term gainer in that process, and China and India, the long-term winners. In any case, the United States will no longer retain its position as the world's sole superpower as a result of this change. The author suggests steps that can be taken to ameliorate the political disturbances that are likely to occur as energy supplies diminish throughout the world.

Laughlin, Robert B. *Powering the Future: How We Will (Eventually) Solve the Energy Crisis and Fuel the Civilization of Tomorrow.* New York: Basic Books, 2011.

The author, cowinner of the 1998 Nobel Prize in Physics, surveys the world as he thinks it might exist two centuries from now, and finds that it probably will have changed far less than most people think it might. The reason for his view is that he believes that a mix of alternative forms of energy—wind, solar, and biomass, for example—will take up the slack left by our diminishing fossil fuel resources.

Linscott, Brad. *Renewable Energy.* Mustang, OK: Tate Publishing, 2011.

The author provides a straight-forward and easy-to-understand description of various forms of alternative energy sources and explains how they can contribute to the solution of the world's energy crisis.

Morriss, Andrew, William T. Bogart, Roger E. Meiners, and Andrew Dorchak. *The False Promise of Green Energy.* Washington, D.C.: The Cato Institute.

This book, published by the conservative think tank the Cato Institute, argues that the promotion of green energy as an alternative to the world's current fossil fuel paradigm is misguided and doomed to failure. The book explains in great detail why efforts to replace coal, oil, and natural gas by wind, tidal, solar, and other power sources will simply be inadequate. The author criticizes government programs for the promotion of alternative energy sources as a huge waste of taxpayer monies.

Nansen, Ralph. *Energy Crisis: Solution from Space*. Burlington, ON: Apogee Books, 2009.

The author has been involved in space engineering for more than four decades. He makes the argument that the world's energy crisis can be solved by making use of and further developing existing solar technologies to replace our dependence on fossil fuels.

Owen, David. *The Conundrum: How Scientific Innovation, Increased Efficiency, and Good Intentions Can Make Our Energy and Climate Problems Worse.* New York: Riverhead Books, 2012.

The author argues that what appear the best-headed and best-intentioned methods of dealing with the Earth's energy crisis may actually work the situation worse because increased efficiency only increasing the demand for products and services, a trend discovered by English economist William Jevons in the late 19th century. The only way to really deal with the crisis, Owen writes, is for individuals to find new and more sustainable ways of living their lives.

Roberts, Paul. *The End of Oil: On the Edge of a Perilous New World.* Boston: Mariner Books, 2005.

The author accepts the premise that the Age of Oil is over and that the world must begin searching for viable options to meet its energy needs. He devotes a considerable portion of the work to an analysis of the economic and political factors associated with the world's dependence on the use of petroleum for a host of purposes. The last part of the book deals with some possible alternative forms of energy, with a discussion of the mechanisms by which the world can move from petroleum-based to alternative-based economies.

Ruppert, Michael C. *Confronting Collapse: The Crisis of Energy and Money in a Post Peak Oil World.* White River Junction, VT: Chelsea Green Publishing, 2009.

Ruppert provides an excellent, classical overview of the argument that the world is rapidly running out of fossil fuel reserves and that nations will soon have to find new ways of powering their economies. He places particular emphasis on the connection between the discovery, recovery, and production of energy resources and national economic policies which, he argues, tends to benefit corporations and small numbers of individuals, at significant cost to the general population. He offers a 25-point presidential plan for making the transition from a fossil-fuel-powered economy to one that no longer depends on oil, natural gas, and coal. The book was the basis for the documentary film *Collapse*, essentially an interview with the author about his position on global energy issues.

Sanchez, Teodoro. *The Hidden Energy Crisis: How Policies Are Failing the World's Poor.* Bourton-on-Dunsmore, Rugby, UK: Practical Action, 2011.

The focus of this book is on the effects of world energy policies on the two billion poorest people across the global. The author argues that those policies have vastly reduced the access of the world's poor to even the simplest kinds of modern energy resources. He recommends a number of policy changes that can provide a more equitable supply of energy resources for the world's poorest populations.

Singer, Clifford E. *Energy and International War: From Babylon to Baghdad and Beyond.* Singapore: World Scientific Publishing Company, 2008.

The author provides an extended and detailed review of armed conflict that has occurred throughout the world's history over access to energy resources, ranging from the Babylonian civilization to the modern-day war in Iraq. He suggests that the changing availability of energy supplies may, counter-intuitively, result in decreased conflict over energy supplies and argues that the Iraq war may be one of the last such conflicts in human history.

Smil, Vaclav. *Energy at the Crossroads: Global Perspectives and Uncertainties.* Cambridge, MA; MIT Press, 2005.

In reviewing the energy issues facing the world today, the author concludes that the most important question is *not* one of supplies and reserves, but, rather, one of the environmental threat posed by the world's dependence on and use of fossil fuels, along with the geopolitical problems posed by the control of fossil fuel reserves by a relatively limited number of nations. He discussed alternative energy sources and lays out some principles by which the transition from fossil fuels to alternative energy sources can come about.

Smil, Vaclav. *Oil: A Beginner's Guide.* London: Oneworld Publications, 2008.

As the title suggests, this book provides a general introduction to basic aspects about the production and consumption of petroleum products. Its five chapters deal with the benefits and burdens of oil, how oil is found, how it is recovered, how it is transported and processed, and how long current supplies are likely to last.

Sovacool, Benjamin K. *Contesting the Future of Nuclear Power: A Critical Global Assessment of Atomic Energy.* Singapore: World Scientific Publishing, 2011.

The author examines the consequences of a new nuclear renaissance for the world economy and environment. He concludes that the harmful results of such a renaissance would far outweigh the benefits, and he argues that societies need to consider more seriously other forms of alternative energy resources in order not to become trapped in the need for additional nuclear resources.

Wulfinghoff, Donald. *Energy Efficiency Manual: For Everyone Who Uses Energy, Pays for Utilities, Controls Energy Usage, Designs and Builds, Is Interested in Energy and Environmental Preservation.* Wheaton, MD: Energy Institute Press, 1999.

This valuable resource provides information about virtually every conceivable aspect of the conservation of energy resources ranging from boiler plant efficiency measurement to fuel oil systems to refrigerant condition to water heating systems to single-duct reheat air systems to personnel door air leakage to lighting layout to an endless list of other topics in the book's more than 1,500 pages.

Yeomans, Matthew. *Oil: A Concise Guide to the Most Important Product on Earth.* New York: The New Press, 2006.

After a brief review of the history of oil as an energy source in human societies, the author focuses on political aspects of the United States' commitment to petroleum during the 20th century. Two chapters are devoted, for example, to detailed discussions of the energy policies of President George W. Bush and the ill-fated CAFE standards of the late 20th century.

Yergin, Daniel. *The Prize: The Epic Quest for Oil, Money, and Power.* New York: Free Press, 2008. *Business Week* magazine called this book "the best history of oil ever written." It provides a fascinating narrative history ranging across virtually every aspect of the growth of petroleum as it becomes the world's primary source of energy for more than a century. It describes the political, geographic, technical, and personal story of the powerful influence of a few important individuals, such as John D. Rockefeller, and the companies they founded, as well as their role in the great political history of the period.

Yergin, Daniel. *The Quest: Energy, Security, and the Remaking of the Modern World.* New York: Penguin Press, 2011.

This book is a follow-up to the author's Pulitzer Prize winning *The Prize.* It begins with the Persian Gulf War of 1990–1991, discusses developments in the role of petroleum in modern society over the following two decades, and speculates as to what the world's energy future might look like.

Periodicals

Balat, M. "A Review of Modern Wind Turbine Technology." *Energy Sources, Part A: Recovery, Utilization, and Environmental Effects* 31, no. 17 (2009): 1561–72.

The author reviews the technical status of wind energy as of 2009, with a consideration as to its possible role in a future alternative energy technology.

Bailis, Robert, and Jennifer Baka. "Constructing Sustainable Biofuels: Governance of the Emerging Biofuel Economy." *Annals of the Association of American Geographers* 101, no. 4 (2011): 827–38.

The rise of interest in biofuels has an alternative source of energy has created new problems as to how such a resource is to be developed and promoted. The authors describe a variety of new mechanisms that go beyond the traditional state as a way of responding to this new challenge.

Beckwith, Robin. "The Tantalizing Promise of Oil Shale." *JPT* 64, no. 1 (2012): 30–36.

This article, in the journal of the Society of Petroleum Engineers, provides an optimistic view of the role that can be played in the world's energy future by oil shale.

Belisle, Peter. "Sustainability/Climate Change." *Strategic Planning for Energy and the Environment* 30, no. 4 (2011): 71–78.

The author reviews efforts in the United States since the 1970s to deal with climate change and depletion of fossil fuel supplies. He discusses the recommendations made by the administration of President Barack Obama, noting that they are "the most aggressive ever proposed in the United States." Still, he concludes, it has been easier to announce goals than to achieve them.

"Britain's Energy Crisis—How Long till the Lights Go Out?" *The Economist* 392, no. 8643 (2009): 9–10.

The author asks "with gas too risky, coal too dirty, nuclear too slow and renewables too unreliable," where is Great Britain going to get the energy it needs to survive and grow in the next few decades. The author has no answer for that question.

Carlvaho, Maria da Graça, Matteo Bonifacio, and Pierre Dechamps. "Building a Low Carbon Society." *Energy* 36, no. 4 (2011): 1842–47.

This paper outlines the strategy adopted by the European Union for dealing with current energy depletion and global climate change issues. The major elements of that program are to reduce greenhouse gas emissions by at least 20 percent, to ensure that 20 percent of final energy consumption is met with renewable sources, and to raise energy efficiency by at least 20 percent by the year 2020.

Cornea, Teodora Manuela Cornea, and Mihai Dima. "Biomass Energy–A Way Towards a Sustainable Future." *Environmental Engineering and Management Journal* 9, no. 10 (2010): 1341–45.

The authors argue that biomass can be an important component in the transition from a dependence on fossil fuels to a reliance on alternative fuels in the coming decades.

Diamond-Smith, Nadia, Kirk R. Smith, and Nuriye Nalan Sahin Hodoglugil. "Climate Change and Population in the Muslim World." *International Journal of Environmental Studies* 68, no. 1 (2011): 1–8.

Although global climate change is likely to have an effect on every country in the world in its own unique way, many parts of the Muslim world are at special risk from such a change. The authors explore some issues of special concern to these nations resulting from climate change.

Friedrichs, Jörg. "Peak Energy and Climate Change: the Double Bind of Post-normal Science." *Futures* 43, no. 4 (2011): 469–77.

The author explores the way in which scientists, national governments, nongovernmental organizations, international agencies, and the general public have attempted (or not attempted) to deal with major coming controversies, such as peak energy and climate change under the circumstances of "post-normal science," which is defined as a situation in which "facts are uncertain, values in dispute, stakes high and decisions urgent."

"The Fukushima Issue." *Bulletin of the Atomic Scientists* 67, no. 4 (2011): 8–22.

This issue of the *Bulletin* includes three articles on "the implications of the Fukushima disaster" for (1) the United States, (2) Europe, and (3) South Korea by, respectively, Mark Cooper, Caroline Jorant, and Soon Heung Chang. The following issue of the magazine, 67, no. 5, devoted the whole issue to a discussion of various aspects of the Fukushima event.

"Future Energy Supply." *Oil and Gas Journal,* July 14, 2003 to August 18, 2003.

This series of six articles discussed a variety of topics relating to peak-energy scenarios including evidence for the existence of peak oil and gas, as well as the technologies that might be able to replace dwindling conventional sources of fossil fuels.

Jonas, Andrew E. G., David Gibbs, and Aidan White. "The New Urban Politics as a Politics of Carbon Control." *Urban Studies* 48, no. 12 (2011): 2537–54.

The authors explore the impact of global climate change and decreasing availability of fossil fuels on the design of modern cities. They say there is a "growing centrality of carbon control" in the determination of urban planning and growth policies.

Liu, L. C., G. Wu, J. N. Wang, and Y. M. Wei. "China's Carbon Emissions from Urban and Rural Households During 1992–2007." *Journal of Cleaner Production* 19, no. 15 (2011): 1754–62.

These researchers examined the contribution of increasing domestic income and expenditure on the contribution of carbon output by individual homes during the period 1992 to 2007, and found that 40 percent of the nation's total carbon output was produced by this source.

Molchanov, Mikhail. "Extractive Technologies and Civic Networks' Fight for Sustainable Development." *The Bulletin of Science, Technology and Society* 31, no. 1 (2011): 55–67.

A battle is developing between transnational corporations involved in the extraction of natural resources and international environmental, labor, and other nongovernmental organizations concerned with sustainable development. The author explores this kind of battle, using the example of the Baku–Tbilisi–Ceyhan oil pipeline in Transcaucasia, a project of an energy conglomerate led by British Petroleum. He uses this issue to draw some conclusions about the resolution of future disputes of this kind in other parts of the world.

Montgomery, Carl T., and Michael B. Smith. "Hydraulic Fracturing: History of an Enduring Technology." *JPT* 62, no. 12 (2010): 26–41.

The authors present an excellent summary of the history of the development and use of hydraulic fracturing, along with some easy-to-understand information about the process itself.

Mulligan, Shane. "Energy and Human Ecology: A Critical Security Approach." *Environmental Politics* 20, no. 5 (2011): 633–49.

Mulligan argues that what he sees as the approaching era of peak oil is likely to become a major security issue for countries around the world that are either exporters or importers of oil. He bases this position on three possible consequences of the appearance of peak oil: uncertainty for the future, threats to freedom, and "the possibility of death."

Ramana, M. V. "Nuclear Power and the Public." *Bulletin of the Atomic Scientists* 67, no. 4 (2011): 43–51.

The nuclear disaster in Fukushima, Japan, in early 2010 once again raised questions about the safety of nuclear power. The author explores the ongoing problems of the nuclear industry in trying to convince the general public that its product is safe, but he points to factors that make this position a continuing "tough sell."

Schnedier, Mycle, Antony Forggatt, and Steve Thomas. "2010–2011 World Nuclear Industry Status Report." *Bulletin of the Atomic Scientists* 67, no. 4 (2011): 60–77.

The authors present a general overview of the status of nuclear power development around the world during the time period mentioned in the title.

Timilsina, G. R., and A. Shrestha. "How Much Hope Should We Have for Biofuels?" *Energy* 36, no. 4 (2011): 2055–69.

The authors offer a review of the increasing role of biofuels in meeting the world's energy needs, but point out a number of problems with depending on this source, one of which is the potential effect it may have on the total food supply available for the world's population.

Victor, David G., and Kassia Yanosek. "The Crisis in Clean Energy." *Foreign Affairs* 90, no. 4 (2011): 112–20.

After a period of very impressive growth, clean energy technologies have begun to experience of reversal of fortunes, largely because of the loss of governmental support for programs that are not yet ready to compete commercially with fossil fuel technologies. This article reviews the forces that led to the development of clean energy technologies and that are now forcing a decline in the field.

"Wonderfuel Gas the World's Energy Crisis Might Be Solved—for Now—by a Plentiful Fuel That's Right under Our Feet." *New Scientist* 2764 (June 12, 2010): 44–47.

This article touts the benefits of natural gas as a replacement for coal and oil and points out that recent explorations have shown that there is an abundant supply of the resource around the world.

Reports

BP Statistical Review of World Energy June 2010. http://www.bp.com/liveassets/bp_internet/globalbp/globalbp_uk_english/reports_and_publications/statistical_energy_review_2008/STAGING/local_assets/2010_downloads/statistical_review_of_world_energy_full_report_2010.pdf. Accessed April 3, 2011.

This report summarizes data on proved reserves, production, and consumption of petroleum, natural gas, and coal in virtually every nation of the world. It is an invaluable resource for information on the current status of fossil fuel statistics throughout the world.

Ebinger, Jane, Walter Vergara, and Irene Leino. *Climate Impacts on Energy Systems: Key Issues for Energy Sector Adaptation.* Washington, D.C.: World Bank Publications, 2011.

Energy production and consumption is widely understood to have significant environment consequences, such as increased levels of air and water pollution and global climate change. Perhaps less well appreciated is the fact that energy production and consumption may themselves be affected by climate change that has already begun to occur and is likely to increase in the future. The authors of this World Bank report review some of these effects and the meaning they will have with respect to needed changes in the way energy is obtained, produced, and used in the future.

The Energy Report: 100% Renewable Energy by 2050. http://assets.wwf.org.uk/downloads/2011_02_02_the_energy_report_full.pdf. Accessed April 3, 2011.

This report was prepared by the World Wildlife Fund, Ecofys, and the Office for Metropolitan Architecture, the latter two, Netherlands-based nongovernmental organizations. The report argues that the data of post-peak petroleum production, along with environmental hazards posed by the world's dependence on fossil fuels, calls for a shift in our energy paradigm from fossil fuels to alternate forms of energy by the middle of this century. It lists 10 ways in which this transformation can be brought about, including a shift to the use of clean energy exclusively, improved use of grids for the distribution of clean energy, bringing an end to energy poverty, investing exclusively in clean and alternate forms of energy, the elimination of food waste, reduction and reuse of materials, and the development of international agreements that promote sustainability and energy conservation.

Grant, Jennifer, Simon Dyer, and Dan Woynillowicz. *Oil Sands Myths: Clearing the Air.* Drayton Valley, Alberta, Canada: The Pembina Institute, June 2009.

This report examines in considerable detail the claims made by government and industry officials about the promise of tar sands as a long-term source of energy and refutes a number of those claims. This report is only one of a number of similar reports issued by the Pembina Institute on the impact of tar sands mining in Alberta.

Hirsch, Robert L., Roger Bezdek, and Robert Wendling. *Peaking of World Oil Production: Impacts, Mitigation, & Risk Management.* [n.p.], February 2005. Available online at http://www.netl.doe.gov/publications/others/pdf/oil_peaking_netl.pdf. Accessed February 7, 2012.

This report was commissioned by the U.S. Department of Energy in order to obtain an assessment of world petroleum supplies; the expected occurrence of oil peaking; and social, political, economic, military, and other consequences of peaking. The 10 conclusions reached by the researchers were that there is no question that peaking is going to occur; the cost of peaking to the U.S. economy will be significant; oil peaking will present a challenge to the nation and the world that is unique in human history; the problem involves liquid fuels (e.g., petroleum), rather than solid (coal) or gaseous (natural gas) fuels; transitioning from petroleum to other fuels will take a significant amount of time; oil peaking involves issues of both supply and demand; solving oil peaking problems is largely a matter of risk management; government intervention in solving oil peaking issues will be essential; economic upheaval is not necessarily a consequence of oil peaking; and more information is needed to understand and solve problems posed by oil peaking.

International Energy Agency. *World Energy Outlook 2010.* http://www.worldenergyoutlook.org/docs/weo2010/WEO2010_ES_English.pdf. Accessed April 3, 2011.

The International Energy Agency (IEA) issues this report annually in November. It summarizes current data on energy production and consumption throughout the world, with projections as to possible future directions for fossil fuels and alternate forms of energy. This website provides the executive summary for the 2010 report, which may be purchased from the IEA at its website, http://www.iea.org/W/bookshop/add.aspx?id = 422. Some topics in the 2010 report include the steps that need to be taken by nations in order to meet the demands of the Copenhagen Accord on global climate change, the effect of emerging economies—especially those of China and India—on future energy consumption patterns, the role that alternate energy can have in the near and more-distant future, and prospects for unconventional forms of petroleum, such as those obtained from oil shale.

International Energy Outlook 2010. Washington, D.C.: U.S. Energy Information Administration, July 2010.

This publication is a treasure trove of information on current and projected production and consumption of the fossil fuels and other energy sources. It contains well over 150 tables, charts, graphs, and figures summarizing information on nearly every aspect of worldwide energy issues.

International Oil Agency. *Oil Market Report.* http://omrpublic.iea.org/omrarchive/18jan11full.pdf. Accessed April 3, 2011.

This report summarizes current production and consumption data for petroleum, and attempts to predict future trends in these areas. As of the date issued (January 18, 2011), data suggest a rise in demand for petroleum products (after a brief drop as the result of worldwide economic downswings), with a continuing decrease in supplies. Authors of the report insert a cautionary note in their introduction to the document: "There may be trouble ahead."

Levine, Mark. *Energy Efficiency in China: Glorious History, Uncertain Future.* Berkeley, CA: Lawrence Berkeley National Laboratory, April 21, 2010.

This report, by the Lawrence Berkeley National Laboratory China Energy Group, provides a fascinating look into the production and consumption of energy in China from 1949 to 2001 and then from 2001 to 2010, followed by some predictions as to what the nation's energy future might look like.

Looking Ahead: Scenarios. Royal Dutch Shell. http://www.shell.com/home/content/aboutshell/our_strategy/shell_global_scenarios/. Accessed April 3, 2011.

Since 1992, Shell has been developing a series of scenarios that attempts to predict possible directions of petroleum production and consumption. These scenarios are used to develop company policy, but they also provide a very useful outlook on a variety of possible petroleum futures for government policy makers and the general public. The most recent of these scenarios, *Signals and Signposts,* was published in February 2011. It described a variety of possible futures depending on the way in which industry, governments, and interested citizens make decisions about how the world's petroleum reserves were to be developed and used.

Oil Depletion Analysis Centre and Post Carbon Institute. *Preparing for Peak Oil: Local Authorities and the Energy Crisis.* [n.p.], [August 2008]. Available online. http://www.odac-info.org/sites/default/files/Preparing_for_Peak_Oil_0.pdf. Accessed February 21, 2012.

The thesis of this report is that national governments and energy corporations have done little to deal with potential consequences of an oil-peaking experience. It remains for individual municipalities, therefore, to begin planning for such an event. This report begins by discussing the evidence for oil peaking and the type of societal changes that may occur as the result of peaking. It then reviews in some detail the types of actions local governments can take in the areas of local finances, airports and highway systems, and food supplies. The report also describes efforts already under way in cities and towns in the United Kingdom and the United States. The Five Principles for Local Officials proposed by one sponsoring organization (The Post Carbon Institute) and the Oil Depletion Protocol are also included in the report.

Rosenfeld, Jaeson, et al. *Averting the Next Energy Crisis: The Demand Challenge.* [n.p.]: McKinsey Global Institute, March 2009.

This report summarizes the McKinsey Global Institute's research on energy demand throughout the world over the short- and long-term future. It concludes that energy demand will continue to decrease modestly in the short term, largely as a result of the worldwide economic downturn, but that it will then begin to increase significantly in most parts of the world. The report provides an abundance of useful statistics on projected global energy demand and points out the potential problems that are likely to arise as worldwide energy supplies begin to fall below demand.

Schneider, Mycle, Antony Froggatt, and Steve Thomas. *Nuclear Power in a Post-Fukushima World.* Washington, D.C.: Worldwatch Institute, 2011.

This report analyzes the effect of the accident at the Fukushima Dai-ichi nuclear facility in March 2011 on the nuclear industry worldwide. It concludes that the industry is unlikely to recover from the disaster and that nuclear power can no longer be considered as a possible long-term replacement for fossil fuels. The authors discuss in some detail the role of renewable energy sources in connection with the nuclear industry and also with the fossil fuel industry itself.

Select Committee on Energy Independence and Global Warming. *Oil Shock: Potential for Crisis.* Washington, D.C.: Government Printing Office, 2010.

This publication provides a transcript of a meeting held before the U.S. House of Representatives Select Committee on Energy Independence and Global Warming, with presentations and commentary by a number of experts in the field.

Sorrell, Steve, et al. *Global Oil Depletion: An Assessment of the Evidence for a Near-term Peak in Global Oil Production.* London: UK Energy Research Centre, August 2009.

The purpose of this report is to answer the question: "What evidence is there to support the proposition that the global supply of 'conventional oil' will be constrained by physical depletion before 2030?" In order to answer that question, the UK Energy Research Centre reviewed more than 500 studies on the current and future supplies of petroleum, along with extensive data obtained from petroleum companies themselves. The summary report is supported by seven detailed technical reports on topics such as "Data Sources and Issues," "Definition and Interpretation of Reserve Estimates," "Decline Rates and Depletion Rates," and "Comparison of Global Supply Forecasts." Researchers concluded that peak oil production is likely to occur before 2030, and may well occur as early as 2020.

U.S. Government Accountability Office. *Crude Oil: Uncertainty about Future Oil Supply Makes It Important to Develop a Strategy for Addressing a Peak and Decline in Oil Production.* Washington, D.C.: Government Accountability Office, February 2007.

The U.S. Governmental Accountability Office conducted this study to determine the answers to three questions: (1) When is peak oil likely to occur in the United States and the rest of the world? (2) What technologies are available to mitigate the effects of peak oil on transportation in the United States? (3) What can federal agencies do to aid in the mitigation of the effects of peak oil on the U.S. economy? The report recommended that the U.S. Department of Energy work with other federal agencies to develop a strategy to "coordinate and prioritize federal agency efforts to reduce uncertainty about the likely timing of a peak and to advise Congress on how best to mitigate consequences."

Who's Winning the Clean Energy Race? 2010 Edition. Pew Charitable Trusts. http://www.pewenvironment.org/uploadedFiles/PEG/Publications/Report/G-20Report-LOWRes-FINAL.pdf. Accessed April 3, 2011.

This report summarizes the progress of research and development on a number of alternate energy technologies, such as solar power, wind energy, and biofuels development. The report includes country-by-country analyses of alternative energy development programs. It concludes that a number of nations, such as China and Germany, are moving forward aggressively in the area of alternate energy development, whereas the United States continues to slip behind other nations in the world.

World Shale Gas Resources: An Initial Assessment of 14 Regions Outside the United States. Washington, D.C.: U.S. Energy Information Administration, April 2011.

This report provides a comprehensive review of shale gas reserves and production in 32 nations in 14 regions outside the United States. The information on these reserves is very detailed and comprehensive.

Nonprint Resources

Ayres, Robert U., and Ed Ayres. "A Bridge to the Renewable Energy Future." *World Watch* 22 (5; September/October 2009).

http://www.worldwatch.org/node/6225. Accessed April 3, 2011.

The authors agree that alternative renewable energy sources will be an essential part of the world's energy future. But such sources will not be widely available for many years. How can nations transition from their current dependence on fossil fuels to alternate fuels sometime in the future? Ayres and Ayres argue that an important bridge during this transition period is the greater use of energy recycling. They provide examples of how this technique is already being used by some corporations to reduce energy consumption.

"The Bakken Formation: How Much Will It Help?" The Oil Drum. http://www.theoildrum.com/node/3868. Accessed November 24, 2011.

This article describes in great detail virtually every aspect of the Bakken Formation, a petroleum- and gas-rich region underlying an extensive part of the central United States. The formation is thought to have some of the largest petroleum and gas reserves in the world, presenting a possible way of delaying peak energy scenarios, although the technology required to extract the resources is not yet sufficient to reach that objective.

Bittle, Scott, and Jean Johnson. "Fast Facts about Energy." http://www.publicagenda.org/whoturnedoutthelights/fastfactsaboutenergy. Accessed April 4, 2011.

The authors of this web page are also the authors of an interesting book on the energy crisis, *Who Turned Out The Lights? Your Guided Tour to the Energy Crisis.* The page consists of 15 graphs that illustrate various points about energy production and use around the world, including production and consumption data, consumption per person data, world oil reserves, sources of electricity, and U.S. carbon dioxide emissions.

Canadian Association of Petroleum Producers. "The Facts on Oil Sands." http://www.capp.ca/UpstreamDialogue/OilSands/Pages/default.aspx#xdDLSsCklUko. Accessed November 23, 2011.

This industry publication does an excellent job of presenting basic factual information about the technology, economics, environmental, and other aspects of tar sands mining, although the presentation is, understandingly, somewhat biased toward an industry view on some controversial topics (such as environmental impacts).

Caruba, Alan. "Alan Caruba Reviews Recent Insights into Peak Oil." anwr.org. http://www.anwr.org/Politics/Alan-Caruba-Reviews-Recent-Insights-into-Peak-Oil.php. Accessed November 29, 2011.

The author presents his argument as to why the concept of peak oil is a hoax perpetrated by "religious fanatics, dictators, and communist thugs who want to line their own pockets, while holding us hostage and enslaving vast portions of the world's population."

Hughes, J. David. "Coal: Some Inconvenient Truths." ASPO-USA. http://www.aspo-usa.org/aspousa4/proceedings/Hughes_David_Coal_ASPOUSA2008.pdf. Accessed November 29, 2011.

This slide show was presented at a meeting of the Association for the Study of Peak Oil—USA on September 23, 2008. It is particularly useful not only because of the excellent analysis of the possible role of coal in the world's energy future, but also because of the extensive collection of data and statistics provided in the program.

"Hydraulic Fracturing Facts." http://www.hydraulicfracturing.com/Pages/information.aspx. Accessed November 29, 2011.

This website is maintained by Chesapeake Energy, a company interested in using hydraulic fracturing in recovering natural gas through the process in the eastern United States. It provides a host of useful information and statistics about the process, although from the standpoint of the energy industry.

Dyer, Simon. "Environmental Impacts of Oil Sands Development in Alberta." http://www.energybulletin.net/node/50186. Accessed November 23, 2011.

This article provides a detailed and extended discussion of the types of environmental damage associated with both strip mined and *in situ* mining of tar sands in the Athabasca region of Alberta, Canada.

Global Energy Crisis. http://global-energy-crisis.com/index.html. Accessed April 4, 2011.

This website provides an excellent introduction to some basic aspects of the world's energy crisis. It discusses at an understandable level the basic scientific facts about the formation of coal, oil, and natural gas; current known resources and reserves; and environmental effects of the combustion of each fuel. It also provides other basic information on units of measurement for energy, future technologies, financial information about production and consumption, and global population trends.

"Global Energy Crisis." *The Economist.* http://www.economist.com/debate/days/view/159. Accessed April 4, 2011.

This website contains one of the ongoing debates on important current social issues sponsored by *The Economist.* In this debate, Joseph J. Romm, of the Centre for American Progress, argues that existing technologies will be adequate for solving the world's energy problems for the foreseeable future. His opponent in the debate, Peter Meisen, of the Global Energy Network Institute, disagrees, arguing that peak oil and global climate change present such serious problems that new and innovative technologies are needed.

"Global Energy Crisis, The." PlanetforLife. http://planetforlife.com/index.html. Accessed April 4, 2011.

PlanetforLife is a personal website hosted by Jack Kisslinger, of Madison, Wisconsin. The website provides access to an extensive and excellent set of resources on virtually all aspects of the global energy crisis, including peak oil, global climate change, sustainable energy, the hydrogen economy, and the 2010 *Deepwater Horizon* blowout.

Heinberg, Richard, and David Fridley. "The End of Cheap Coal." *Nature* 468 (7322; November 18, 2010). http://www.nature.com/nature/journal/v468/n7322/full/468367a.html (subscription required). Accessed April 4, 2011.

The authors explain why future global energy policies built on a reliance on coal are incorrect and inadequate since the rising price of coal production will make that fuel too expensive for widespread use in the future.

Helms, Lynn. "Horizontal Drilling." https://www.dmr.nd.gov/ndgs/newsletter/NL0308/pdfs/Horizontal.pdf. Accessed November 22, 2011.

This article provides a clear explanation of a relatively new process for extracting petroleum and natural gas from reserves by drilling horizontally, rather than vertically, into a reservoir. The technology holds the promise for vastly increasing the quantities of oil and natural gas that can be recovered from known fields, thus extending significantly into the future possible peak oil and peak gas scenarios in many parts of the world.

Hirsch, Robert L. "The Inevitable Peaking of World Oil Production." The Atlantic Council of the United States. 16 (3; October 2005). http://www.acus.org/docs/051007-Hirsch_World_Oil_Production.pdf. Accessed April 3, 2011.

This article provides a very nice overall summary of the coming decline of oil production in the world, along with its potential effects on human societies. The author points out that a number of replacements are already available for petroleum—such as oil sands, coal liquefaction, gas-to-liquid technologies, and conservation—although there has been inadequate development of these procedures for general use. He also explains why a petroleum crisis is different from an energy crisis, in that petroleum can be replaced for many of its current applications by alternative fuels, although its use in transportation devices can not.

"Horizontal and Multilateral Wells." JPT Online. http://www.spe.org/spe-app/spe/jpt/1999/07/frontiers_horiz_multilateral.htm. Accessed November 23, 2011.

This article presents a clear explanation of the technologies involved in horizontal and multilateral oil well drilling, with an analysis of the relative costs of these two technologies compared to conventional vertical oil drilling.

Hubbert, M. King. "Nuclear Energy and the Fossil Fuels." Post Carbon Institute. http://www.energybulletin.net/node/13630. Accessed April 2, 2011.

This article is one of the most famous documents in the history of energy studies. It was first read by Hubbert at a meeting of the American Petroleum Institute in March 1956, and then published in print form as Publication No. 95 by the Shell Development Corporation in June 1956 and *Drilling and Production Practice* in the same year. It was reprinted in *Energy Bulletin* on March 8, 2006. The article is of significance because of Hubbert's contention that production of petroleum in the United States would peak in the 1970s and then begin to decline there and around the world. Although Hubbert's timeline is obviously incorrect, his concept of a peak period for the production of petroleum (and other fossil fuels) was a revolutionary and disturbing concept for his

1956 audience, as it continues to be for many observers today.

Jeffries, Elisabeth. "Energy Efficiency, Rediscovered: Climate Change and Rising Energy Prices Are Making Efficiency Look Good—Again. *World Watch.* 22 (1; January 2009). http://www.worldwatch.org/node/5968. Accessed April 3, 2011.

The author discusses in detail the increasing emphasis on energy conservation seen in many nations, sometimes promoted by energy producers themselves, as a way of dealing with decreasing energy supplies.

Klare, Michael T. "The End of the World As You Know It." Countercurrents.org. http://www.countercurrents.org/klare160408.htm. Accessed April 3, 2011.

The author discusses five ways in which the world is likely to experience dramatic changes as the result of diminishing supplies of the fossil fuels. They are greater competition among nations for decreasing energy supplies; a growing insufficiency of primary energy resources; the painfully slow development of alternative energy sources; a steady movement of financial wealth from energy-poor to energy-rich nations; and an increased risk of armed conflict over energy resources.

Klare, Michael T. "The Permanent Energy Crisis." TomPaine.com. http://www.tompaine.com/articles/2006/02/13/the_permanent_energy_crisis.php. Accessed April 4, 2011.

This article is important because it puts forward in clear terms the major components of the so-called world energy crisis. Klare discusses the problems associated with the simultaneous rise in demand and decrease in production of petroleum and some possible ways of dealing with such problems.

Lendman, Stephan. "Peak Oil—True or False?" Global-Researcher.ca. http://www.globalresearch.ca/index.php?context = va&aid = 8260. Accessed November 21, 2011.

This very well-researched and written article reviews the nature of the peak oil argument and presents statements for and against the accuracy of this theory of petroleum supplies and production.

MacDonald, Gregor. "Will Global Warming Amplify The Energy Crisis? Look At Australia." Business Insider. http://www.businessinsider.com/china-lights-global-floods-australian-coal-2011-1. Accessed April 4, 2011.

As petroleum resources throughout the world become more and more inadequate for dealing with growing energy demands, many experts expect that coal will begin to replace oil as the world's primary energy resources. The potentially devastating environmental effects arising out of a shift to coal, however, can be significant, as outlined in this article.

Mortished, Carl. "Energy Crisis Cannot Be Solved by Renewables, Oil Chiefs Say." *The Times,* June 25, 2007. http://business.timesonline.co.uk/tol/business/industry_sectors/natural_resources/article1980407.ece. Accessed April 4, 2011.

Leaders of international oil companies warn the general public that hopes that green energy will be able to meet the world's energy needs in the future are misplaced. Only fossil fuels are available in quantities sufficient to meet future demands, they say. One spokesman, Jeroen van der Veer, chief executive officer of Royal Dutch Shell, says that "there is no shortage of oil and gas in the ground," but companies have not yet developed the technology to access those resources.

"Natural Gas Advances Save The World From Energy Crisis." Oilprice.com. http://oilprice.com/Energy/Natural-Gas/Natural-Gas-Advances-Save-The-World-From-Energy-Crisis.html. Accessed April 3, 2011.

This article, published on December 12, 2009, claims that the so-called energy crisis "may very well be over." The reason for this optimistic assessment is that engineers have apparently developed an economically viable method for extracting liquid petroleum from oil shales. Oil shales are so widely abundant throughout the world, according to the article, that the world now appears to be "much further away from running out of energy any time soon."

Offshore. http://www.offshore-mag.com/index.html. Accessed November 29, 2011.

Offshore is a journal that provides technical and general information on essentially every aspect of the offshore oil and gas drilling industry. Although designed primarily for professionals who work in the field, it and its website provides probably the most complete set of data and information about not only technical procedures, but also important issues relating to offshore drilling.

Roberts, Paul. "Got Gas? Oil Isn't the Only Fossil Fuel That's in Crisis." Slate.com. http://www.slate.com/id/2100318/. Accessed April 4, 2011.

Roberts points out that concern about the world's energy crisis generally refer to the imbalance in production and consumption of petroleum. But, he says, there is a similar problem with natural gas, in that, especially as supplies of petroleum become unequal to demands for the fuel, a consequent deficiency in natural gas supplies is also likely.

Toro, Francesco. "The Red Apertura and the Forgotten Art of Disarming Atomic Bombs." Caracas Chronicles. http://caracaschronicles.com/2011/09/26/the-red-apertura-and-the-art-of-disarming-atomic-bombs/. Accessed November 23, 2011.

Francesco provides a detailed and very illuminating review as to why Venezuela has made so little progress in developing its Orinoco Belt, with the world's largest reserves of extra-heavy crude (or any other type of petroleum product) in the world.

"World Energy Supply." Earth. http://www.theglobaleducation-project.org/earth/energy-supply.php. Accessed April 4, 2011.

Planet Earth is a project of educators in British Columbia. Its website provides a vast amount of data on a variety of environmentally important topics. Much of these data are presented in the forms of charts, graphs, and tables. The section on Energy Supply, for example, contains visuals on proven crude oil reserves, proven coal reserves, proven natural gas reserves, major oil producing nations, and similar topics.

Yergin, Daniel. "There Will Be Oil." The Wall Street Journal Saturday Essay. http://online.wsj.com/article/SB1000142405 3111904060604576572552998674340.html. Accessed November 21, 2011.

Yergin is one of the best known and most articulate critics of the theory of peak oil. He argues in this essay that predictions of peak oil are incorrect because they use incorrect estimates of the amount of oil still available in the Earth and that future exploration will result in the discovery of vast new deposits of the fossil fuel.

Ziagos, John, and Ken Wedel. "Energy Crisis: Will Technology Save Us?" Science on Saturday 2007. http://www.llnlretirees.org/PDFfiles/2008-04-16_Ziagos_SOS07_energy.pdf. Accessed April 3, 2011.

Science on Saturday is a series of lectures by researchers and teachers on important topics in the field of science. This slide show provides a superb general introduction to the coming world energy crisis, with background in the science involved and up-to-date data on the production and use of various forms of energy.

7 Chronology

Energy has been an integral part of human civilization since the earliest days. Whenever a person cooks a meal, heats a home, removes water from a river or a lake, moves heavy objects from one place to another, or performs some other kind of work, he or she makes use of some form of energy. Of course, the forms of energy available to humans evolved and became more complex over time. As they did so, they began to have their own impact on human civilization, particularly on human health, as well as on the natural environment. This chapter summarizes some of the major events that have occurred throughout human history involving important energy developments.

ca. 400,000 BCE Estimated first controlled use of fire by humans.

ca. 8000 BCE Estimated first use of oil lamps by humans.

ca. 6000 BCE Estimated first use of domesticated animals for use in transportation. The first animals to be domesticated for this purpose were probably the camel, the horse, the ox, and the ass.

ca. 5000–6000 BCE First use of rock oil (asphalt, bitumen, or other forms of petroleum) as an adhesive, a filler, a medicine, and a fuel in lamps, primarily in Sumeria, Mesopotamia, and Babylonia.

With an offshore oil platform in the distance, a group of students play in the water in Santa Barbara, California. (AP Photo/Michael A. Mariant)

ca. 3000 BCE First use of coal as a fuel by the Chinese.

ca. 2000 BCE Evidence for the first use of horse-drawn vehicles in Egypt and China.

ca. 1000 BCE The Temple of Delphi is built in Greece at the site of a natural gas seep in the ground. The gas is used to keep the temple fire burning, a practice that probably existed for hundreds of years before its use at Delphi.

ca. 500 BCE Greeks make use of solar energy to heat their homes during the winter.

ca. 500 BCE The Chinese develop methods for collecting natural gas through bamboo pipes for use in heating and other purposes.

ca. 380 BCE Greece is, for all practical purposes, denuded of forests because of the demand for trees for use in heating, building, industrial operations, and other uses. The harvesting and use of wood is severely restricted by the government.

ca. 325 BCE Greek natural philosopher Aristotle provides the first definition of energy (*energeia*) as, in modern terms, "being at work."

ca. 315 BCE Greek natural philosopher Theophrastus writes the first formal text on minerals in which he describes the properties of a black stone that burns (coal).

ca. 240 BCE Greek inventor Ctesibius of Alexandria invents the double action piston pump, one of the first machines for doing work, used to lift water from one level to another.

ca. 200–400 BCE Mention of the use of water to power water wheels is common in European and Asian literature, although the use of such devices probably goes back much earlier in history.

ca. 200 BCE Chinese rulers issue a series of laws that may be regarded as the earliest environmental regulations, limiting the harvesting and burning of wood during certain parts of the year.

ca. 120 BCE The first mention of coal in Chinese literature occurs in the writings of Liu An (died 122 BCE). By the time of Liu's writings, coal is widely used for heating in Chinese homes.

ca. 80 BCE Greek poet Antipater of Thessalonika tells of the use of water power to replace the manual labor formerly performed by young girls.

47 CE Roman natural philosopher Pliny the Elder writes of the use of "dried mud," a type of peat, by the Chauci tribe for heating their homes, the first written record of the use of peat as a fuel.

ca. 50 Greek inventor Heron (Hero) of Alexandria constructs the first steam engine and the first wind-powered engine, a primitive windmill.

ca. 100 Roman occupiers of Great Britain use surface-mined coal for heating and industrial purposes. Although native tribes probably used coal for the same purposes, no written records of the use of coal by native British peoples prior to the 13th century exist.

347 Chinese workers collect oil from wells as deep as 800 feet using bamboo poles with metal tips.

534 The Justinian code establishes sun rights for all Romans, declaring that a structure may not deprive a resident of access to sunlight in his or her home.

787 First written records of tidal power being harnessed for useful purposes off the northeastern coast of Ireland.

1264 Marco Polo writes of seeing the residents of Baku (modern Azerbaijan) collecting oil from pools that have seeped to the surface for use as a heating and a lighting material. Residents explain that they have used this practice for many centuries.

1272/1273 King Edward I announces the first air pollution law in history, declaring it illegal to burn sea coal in the city of

London. The law is needed because pollution from coal smoke is so bad as to pose a serious health risk to residents of the city. Edward follows this proclamation with other, more severe, restrictions on the burning of coal in the city.

1330 Residents of the French town of Chaudes-Aigues use water heated by geothermal energy for heating their homes. Geothermal energy is still used for heating and industrial operations in Chaudes-Aigues today.

1379 In one of the first efforts to control air pollution, the English government assesses a tax on the purchase of coal in hopes of discouraging use of the fuel.

1482 King Edward IV establishes the barrel as a unit of measurement equivalent to 42 gallons for the sale of herring. The unit eventually becomes more widely used for a variety of products and materials, one of which is crude oil. Oil production and consumption is still commonly expressed in units of barrels (bbl).

1594 Persian engineers dig the first true oil well in the region of Baku, in present-day Azerbaijan.

1594 Dutch mill owner Cornelis Corneliszoon van Uitgeest constructs a saw mill operated by wind power in the town of Uitgeest. The mill floated on a raft that could be rotated to take maximum advantage of prevailing winds.

ca. 1600 Overwhelming demands for wood for heating, production of charcoal, construction of ships, and other purposes lead to the destruction of most forests in Great Britain. To replace wood for heating and industrial operations, the British begin to increase their reliance on coal, of which there are abundant supplies.

1603 Sir Hugh Platt, a writer on food and food technology, suggests that coal can be heated to drive off volatile components, leaving behind a solid material, coke, which burns with greater efficiency than does coal itself. Coke eventually becomes one of the most important fuels in industrial operations, such as the production of steel.

1615 French engineer and physicist Salomon de Caux (Caus) constructs a solar-powered water pump that uses a magnifying glass to focus the sun's rays on a metal sphere containing water. Steam produced within the sphere is used to do work. The invention appears to be the first use of solar power for practical applications since the days of classical Greece.

1661 English writer John Evelyn publishes a pamphlet titled "Fumifugium: or The Inconveniencie of the aer and smoak of London dissipated," probably the world's first critique of the nature and hazards of severe air pollution.

1667 English adventurer Sir Thomas Shirley describes the flow of a flammable gas from a coal deposit near Wigan, England. This report is the earliest mention of natural gas in Europe.

1673 French explorer Louis Joliet makes the first record of coal deposits in the United States, along the Illinois River in northern Illinois.

1684 English minister John Clayton discovers that a flammable gas is produced when coal is heated in the absence of air. Clayton collects this coal gas in a rubber bladder, but does not suggest any practical use for the product.

1690 French inventor Denis Papin builds the prototype of a steam engine that he suggests can be used to pump water from a mine. Papin earlier invented a steam digester, similar to the modern-day pressure cooker.

1698 English military engineer and inventor Thomas Savery obtains a patent for an improved version of Papin's steam engine (1690) for use in removing water from mines.

1712 English inventor Thomas Newcomen invents the first successful atmospheric steam engine.

1719 A Cree trader named Waupisoo brings a sample of tar sands to a trading post at Fort Churchill, the first mention of tar sands in history.

1748 The first commercial coal mine begins operation in the United States near Richmond, Virginia.

1762 Anthracite (hard) coal is discovered in the Wyoming Valley of northeastern Pennsylvania. Experts estimate that more than 16 billion tons of anthracite are buried in the region, setting off one of the largest searches for coal anywhere in the world.

1766 English chemist Henry Cavendish discovers hydrogen.

1769 English inventor James Watt devises an improved version of Thomas Newcomen's steam engine.

1769 French inventor Nicolas-Joseph Cugnot builds what is probably the world's first self-propelled vehicle, or automobile. The car is powered by a steam engine that must be recharged ever few minutes, making it essentially impractical. In 1771, the vehicle collided with a stone wall, making it the first automobile accident in history.

1775 French inventor Jacques Perrier invents a ship powered by steam engines.

1775 Ten-year-old Pierre-Simon Girard invents the water turbine, which in modified forms is a critical element in the production of electricity from running water.

1791 English inventor John Barber obtains a patent for the first gas turbine engine. Because of inadequate technology, the engine is never actually built, although Barber's concept for its design was sound.

1799 Italian physicist Alessandro Volta constructs the first battery, which consists of a series of two different metals separated by moist paper.

1799 A French father-and-son team by the name of Girard file a patent application for a device that uses the energy of ocean waves to power saws, pumps, and other types of machinery.

1800 At the beginning of the Industrial Revolution, the concentration of carbon dioxide in the Earth's atmosphere is measured at 283 ppm (parts per million). As of June 2011, that number had risen to 394 ppm, an increase of 39.2 percent in 111 years.

1801 English inventor Richard Trevithick builds the first steam locomotive.

1807 French politician and inventor François Isaac de Rivaz invents the world's first internal combustion engine, designed to operate an automobile by burning a mixture of oxygen and hydrogen. The invention is not a commercial success.

1810 The first reported coal mine explosion in the United States occurs in Virginia. There are no deaths in the accident.

1812 German entrepreneur obtains a charter for the London and Westminster Gas Light and Coke Company for the purpose of producing and selling coal gas to be used as a source of illumination in the city. The company is the first public gas lighting system in Great Britain.

1820 Reverend William Cecil, at Magdalene College, Cambridge, invents an engine powered by the combustion of hydrogen gas.

1821 American entrepreneur William Hart, sometimes called the Father of Natural Gas in the United States, drills the first natural gas well in Fredonia, New York. The gas is used to light homes in the city and to cook food at a local hotel.

1828–1842 Inventors in Hungary, Scotland, the Netherlands, and the United States propose a variety of designs for an automobile powered by electricity.

ca. 1830 English chemist and physicist Michael Faraday and American physicist Joseph Henry independently discover the principle of electromagnetic induction, which forms the basis of both electric generators and electric motors.

1840 First recorded use of natural gas for a commercial project, for the evaporation of brine to produce salt at Centerville, Pennsylvania.

1842 English jurist and physical scientist Sir William Robert Grove invents the first fuel cell, in which hydrogen and oxygen are combined to form water and energy.

1848 Russian engineer F. N. Semyenov digs what is the world's first true oil well on the Aspheron peninsula near Baku. By 1861, the well is producing about 90 percent of the world's supply of crude oil.

1854 In an effort to find an inexpensive substitute for whale oil in lamps, Polish druggist Ignacy Lukasiewicz digs his own oil well and builds a distilling tower to extract kerosene from crude oil.

1857 The world's first large oil refinery opens in Cîmpina, Romania.

1858 Explorers dig the first oil wells in North America in southern Ontario near the town of Petrolia.

1859 Colonel Edwin Drake drills the first successful oil well in the United States at Titusville, Pennsylvania.

1860 French mathematics teacher and inventor Augustine Mouchot invents the first system for using solar power to operate machinery.

1866 The first commercial coal strip mine in the United States opens near Danville, Illinois.

1870 John D. Rockefeller founds Standard Oil of Ohio. In less than a decade, the company has more than 90 percent of petroleum-refining capacity in the United States.

1876 English professor of natural philosophy William Grylls Adams and his student Richard Evans Day devise a method for producing electricity from sunlight using a primitive form of a solar cell.

1878 German engineer Sigmund Schuckert builds the world's first electric power station in Ettal, Bavaria. Electricity is produced from steam generated by the burning of coal.

1882 American inventor Thomas Edison opens the world's first commercial electric power station in New York City.

1882 American paper manufacturer H. J. Rogers builds the world's first hydroelectric power plant on the Fox River, in Wisconsin.

1883 G. C. Hoffman, of the Geological Survey of Canada, for the first time successfully separates bitumen from oil sands.

1885 Oil is discovered in Sumatra, Indonesia.

1887 Scottish inventor James Blyth constructs the first windmill for the production of electricity.

1890 The U.S. Congress passes the Sherman Antitrust Act.

1891 The first offshore oil wells in the world are drilled in Grand Lake St. Marys, Ohio.

1891 The first natural gas pipeline is built from a well source in central Indiana to the city of Chicago, a distance of about 120 miles.

1896 The first saltwater offshore oil well is drilled off a pier extending into the Santa Barbara Channel near Summerland, California.

1903 Entrepreneur Henry Ford establishes the Ford Motor Company, initiating an era in which the automobile becomes widely available to the general public.

1906 The federal government files suit against Standard Oil of Ohio under the Sherman Antitrust Act.

1908 Oil is discovered at Masjid-i-Suleiman, Persia (now Iran), by British explorer George Reynolds.

1910 The U.S. government authorizes the creation of land for use as Naval Petroleum Reserves.

1910 Oil is discovered in Mexico.

1911 The U.S. Supreme Court orders the breakup of Standard Oil.

1911 The world's first geothermal power generating plant is built in Larderello, Italy. Electricity produced at the plant is used to power the regional electrical railway system.

1924 The U.S. Congress passes the Oil Pollution Control Act, the first legislative action in the United States designed to reduce air pollution from the use of petroleum.

1930 Entrepreneur Robert Fitzsimmons makes the first commercial sale of bitumen produced by tar sands. The product is used as a fence post dip, for roof tar, and as a paving material.

1932 Oil is discovered in Bahrain.

1933 The government of Saudi Arabia signs an agreement with Standard Oil, allowing them to drill for oil. Five years later, the first producing wells began to operate in the nation.

1934 The world's first offshore drilling rig begins producing oil at Artem Island, Azerbaijan.

1935 Du Pont scientist Wallace Carothers invents nylon, the first completely synthetic fiber. The invention is important because it opens a whole new use for petroleum, from which nylon and many other synthetic products are made.

1938 Mexico nationalizes all oil companies, organized under a single entity, Pemex.

1938 Oil is discovered in Kuwait and Saudi Arabia.

1942 Working at the University of Chicago, researchers produce the first controlled nuclear fission reaction, a reaction that forms the basis of all future nuclear power plants.

1948 The Ghawar field, by far the largest oil field in the world, is discovered in Saudi Arabia.

1951 Iran nationalizes the Anglo Iranian Oil Company.

1956 Oil is discovered in Algeria and Nigeria.

1956 The world's first commercial nuclear power plant, Calder Hall, opens at Sellafield, England.

1960 Iran, Iraq, Kuwait, Saudi Arabia, and Venezuela join to form the Organization of the Petroleum Exporting Countries (OPEC) for the purpose of negotiating with oil companies for the production of petroleum in their nations.

1967 The oil tanker *Torrey Canyon* runs aground on Seven Stones reef in Scotland, dumping 120,000 tons of crude oil into the sea before sinking. The event was the first major oil disaster in history.

1967 Sunoco, Inc., opens the Great Canadian Oil Sands Plant in northern Alberta, Canada, with the goal of extracting 300 billion barrels of petroleum from the Athabasca oil sands field.

1968 A consortium of oil companies led by British Petroleum (BP) discovers one of the largest oil fields in North America in Alaska's Prudhoe Bay.

1970 Oil production peaks in the United States. Oil production also peaks in Venezuela and Libya in the same year.

1973 OPEC declares an oil embargo against nations that supported Israel in the Yom Kippur war of that year. The embargo lasts until March 1974.

1979 Following a revolution that replaces a pro-Western with an Islamic government, Iran dramatically curtails the sale of oil to most western nations, especially the United States, an act that President Jimmy Carter calls "the moral equivalent of war."

1980 The world's first wind farm, consisting of 20 30 kW turbines, is opened in southern New Hampshire.

1990 The United States and allied forces initiate Operation Desert Storm, a campaign to beat back the armed forces of Iraq who had attacked Kuwait earlier in the year. During their retreat, Iraqi forces destroyed in excess of 700 oil wells with a loss of between 46 and 138 million tons of oil.

1991 The first offshore wind farm is created at Vindeby, Denmark, consisting of 11 450 kW turbines.

1998 The world market price of oil stands at about $10 a barrel.

2003 The administration of President George W. Bush introduces "Clear Skies" legislation amending and weakening the Clean Air Act.

2005 The vast majority of the world's nations ratify the Kyoto Protocol to reduce greenhouse gas emissions. The United States is one of the few nations not to ratify the treaty.

2008 The price of oil reaches $147.27 per barrel, the highest price ever for oil. Less than six months later, oil prices drop to about $34 per barrel.

2009 The largest single oil development in history, the Khurais oilfield in Saudi Arabia, is brought on stream.

2010 Explosion at BP's *Deep Water Horizon* drilling rig in the Gulf of Mexico results in one of the world's greatest oil-related environmental disasters.

2011 A 9.0 magnitude earthquake and resulting tsunami strike the coast of Japan, damaging a number of nuclear reactors and raising questions about the safety of nuclear power generation.

Glossary

Discussions of energy-related issues frequently involve a number of technical terms. This glossary is offered to provide a better understanding of some of those terms.

alcohol See ethanol.

alternative energy source An energy source other than a fossil fuel, such as wind energy, solar power, or nuclear power.

anthracite A form of coal, also known as "hard coal," with the highest portion of pure carbon of any type of coal.

biodiesel A type of fuel made from canola, soybean, or other vegetable oils; animal fats; or recycled grease used as a substitute for petroleum-based fuels.

biomass Any material made from organic matter, such as grass or leaves.

biomass gas A fuel produced by the action of microorganisms on biomass, consisting primarily of methane and carbon dioxide.

bitumen A black, viscous liquid mixture that remains after the distillation of crude oil.

bituminous coal An intermediate grade of coal between anthracite and lignite coals. Also known as soft coal.

breeder reactor A type of nuclear reactor that produces as one by-product isotopes that can then be used for other fission reactions.

carbon dioxide (CO_2) A colorless, odorless substance produced during the combustion of any organic material, including all fossil fuels.

coal gassification process of converting coal into gas.

coal liquefaction The process of converting coal into a liquid.

coke A high energy fuel produced by heating soft coal in the absence of air.

Corporate Average Fuel Economy (CAFE) A set of standards established by the U.S. government dictating the fuel efficiency required of new cars built and sold in the United States.

crude oil A mixture of hydrocarbons that exists in liquid phase in natural underground reservoirs.

E85/E95 A fuel containing either 85 percent ethanol and 15 percent gasoline or 95 percent ethanol and 5 percent gasoline, respectively.

energy efficiency The ratio of useful work obtained compared to the energy input in a system.

ethanol An organic compound with the chemical formula C_2H_5OH that can be used by itself or in combination with other substances as fuel.

exploratory well well that is dug to determine whether useable supplies of oil and/or natural gas are available in an area.

fission A nuclear reaction in which a large atomic nucleus, such as that of a uranium atom, is split into smaller parts, with the release of energy.

fly ash Particulate matter consisting of particles less than 10–4 m in diameter produced during the combustion of coal.

fracking (also hydrofracking or hydraulic fracturing): A process for extracting oil and natural gas from a rocky reservoir by splitting the rock by injecting a liquid under high pressure.

global climate change Long-term changes in weather patterns that may be associated, among other factors, with the release of carbon dioxide into the Earth's atmosphere.

hydraulic fracturing *See* fracking.

lignite The lowest type of coal in terms of carbon content and heat content. It is used almost exclusively for use in electric power generating plants.

liquefied petroleum gas (LPG) mixture of hydrocarbons obtained during the fractionation of petroleum or natural gas, which is then liquefied for easier shipping.

nitrogen oxides (NO*x*) Compounds of nitrogen and oxygen, such as nitrous oxide (N_2O), nitric oxide (NO), and nitrogen dioxide (NO_2), commonly formed during the combustion of fossil fuels.

"number 00" fuel oil Six categories of fuel oil, 1 to 6, established by the American Society for Testing and Materials that indicate various uses for fuel oils of different characteristics.

ocean thermal energy conversion (OTEC) Any technology for producing energy by harnessing temperature differences between surface and deep ocean waters.

oil shale A sedimentary rock that contains kerogen, a solid organic material.

passive solar heating Any solar heating system that uses no external mechanical power for the collection of solar power.

peak oil/coal/natural gas The time at which the maximum rate of oil, coal, or natural gas has been reached, after which production begins to decline for that fuel.

petrochemical A substance obtained from petroleum or natural gas, often used for a very wide variety of synthetic commercial products.

petroleum A broad term for mixtures of solid, liquid, and gaseous hydrocarbons present in crude oil and in various refined forms of crude oil; sometimes used synonymously with the term crude oil.

photovoltaic cell An electric cell used to convert solar energy into electrical energy.

probable energy reserves Estimated amounts of an energy resource that can reasonably be expected to exist and to be recoverable under existing and technical conditions.

proved energy reserves Estimated amounts of an energy resource that, with reasonable certainty, are recoverable under existing economic and technical conditions.

recoverable reserves extent to which a fossil fuel can be extracted from a given source, given all relevant technological and economic conditions.

refinery A plant used to separate crude oil into various constituents, such as kerosene, gasoline, and fuel oils, based on the temperatures at which they boil.

renewable energy resources Energy resources that replenish naturally, such as biomass, hydroelectric power, geothermal, solar, wind, ocean thermal, wave, and tidal.

shaft mine An underground mine that is reached by means of a vertical shaft.

shale gas Natural gas that is produced from oil shale formations.

solar constant The average amount of solar radiation that reaches the Earth's upper atmosphere on a surface perpendicular to the sun's rays; equal to 1353 watts per square meter or 492 Btu per square foot.

speculative resources Undiscovered coal, oil, or natural gas whose existence is suspected because of secondary evidence, such as their presence in an existing energy resource field.

steam turbine A mechanical device used to convert thermal energy of steam into electrical energy.

strip mine A mine in which rocky and other overburden is first removed, allowing direct access to a natural resource, such as coal.

tailings The waste that remains after a metal-bearing ore has been removed. Tailings consist of finely ground rock and chemicals used in the extraction process.

tar sands Naturally occurring sands that contain bitumen and that can be processed to obtain mixtures of liquid hydrocarbon which, in turn, can be converted into finished petroleum products.

vertical integration A situation in which a company controls more than one stage of production and distribution, as in the case of a coal company that also owns a railroad used to transport coal.

waste heat recovery Any system that captures heat produced in an industrial operation and uses that heat from some useful purpose.

Index

About the Author

David E. Newton holds an associate's degree in science from Grand Rapids (Michigan) Junior College, a BA in chemistry (with high distinction) and an MA in education from the University of Michigan, and an EdD in science education from Harvard University. He is the author of more than 400 textbooks, encyclopedias, resource books, research manuals, laboratory manuals, trade books, and other educational materials. He taught mathematics, chemistry, and physical science in Grand Rapids, Michigan, for 13 years; was professor of chemistry and physics at Salem State College in Massachusetts for 15 years; and was adjunct professor in the College of Professional Studies at the University of San Francisco for 10 years. Previous books for ABC-CLIO include *Global Warming* (1993), *Gay and Lesbian Rights—A Resource Handbook* (1994, 2009), *The Ozone Dilemma* (1995), *Violence and the Mass Media* (1996), *Environmental Justice* (1996, 2009), *Encyclopedia of Cryptology* (1997), *Social Issues in Science and Technology: An Encyclopedia* (1999), *DNA Technology* (2009), and *Sexual Health* (2010). Other recent books include *Physics: Oryx Frontiers of Science Series* (2000), *Sick!* (4 volumes) (2000), *Science, Technology, and Society: The Impact of Science in the 19th Century* (2 volumes; 2001), *Encyclopedia of Fire* (2002), *Molecular Nanotechnology: Oryx Frontiers of Science Series* (2002), *Encyclopedia of Water* (2003), *Encyclopedia of Air* (2004), *The New Chemistry* (6 volumes; 2007), *Nuclear Power* (2005), *Stem Cell Research* (2006), *Latinos in the Sciences, Math, and Professions* (2007), and *DNA Evidence and Forensic Science* (2008). He has also been an updating and consulting editor on a number of books and reference works, including *Chemical Compounds* (2005), *Chemical Elements* (2006), *Encyclopedia of Endangered Species* (2006), *World of Mathematics* (2006), *World of Chemistry* (2006), *World of Health* (2006), *UXL Encyclopedia of Science* (2007), *Alternative Medicine* (2008), *Grzimek's Animal Life Encyclopedia* (2009), *Community Health* (2009), and *Genetic Medicine* (2009).